P9-CMX-675

EXTRA SENSORY

ALSO BY BRIAN CLEGG

Gravity

The Universe Inside You

How to Build a Time Machine

Inflight Science

Armageddon Science

Before the Big Bang

Ecologic

The God Effect

A Brief History of Infinity

The Man Who Stopped Time

The First Scientist

Light Years

EXTRA SENSORY

THE SCIENCE AND PSEUDOSCIENCE OF TELEPATHY AND OTHER POWERS OF THE MIND

BRIAN CLEGG

St. Martin's Press *New York*

Printed in the United States of America. For information, address St. Martin's Press, 175 Fifth Avenue, New York, N.Y. 10010.

www.stmartins.com

ISBN 978-1-250-01906-6 (hardcover)
ISBN 978-1-250-03168-6 (e-book)

St. Martin's Press books may be purchased for educational, business, or promotional use. For information on bulk purchases, please contact Macmillan Corporate and Premium Sales Department at 1-800-221-7945 extension 5442 or write specialmarkets@macmillan.com.

First Edition: May 2013

10 9 8 7 6 5 4 3 2 1

To Gillian, Chelsea, and Rebecca

CONTENTS

ACKNOWLEDGMENTS

My thanks to everyone at St. Martin's Press who have put so much work into this book, particularly my editor, Michael Homler. A real debt is owed to the many researchers described in the book who have been prepared to risk their reputations to investigate a subject that many scientists consider untouchable.

EXTRA SENSORY

1.

FEEL THE POWER—SUPERHEROES AND PHYSICS

Admit it. At some time in your life you have probably wished that you had superpowers. There is something particularly appealing about being special, having something above and outside the ordinary, being able to go far beyond the conventional capabilities of a human being. As a kid, I certainly did a fair amount of running around with one arm stuck out in front of me, playing Superman, pretending that I could fly.

Even as an adult, the idea of having such paranormal abilities appeals. You may try to conceal the delight in a cloud of sophistication, but you have to be totally lacking in imagination not to get a frisson of excitement at the thought of having such talents—and the continuing box-office success of superhero movies shows that the public at large can't get enough of the vicarious enjoyment of the feats that are possible with superpowers.

Unfortunately, when we drop the suspension of disbelief required to enjoy such a movie and examine superheroes through the scientific microscope, most superpowers fall away like discarded spiderwebs. They could never be real, as they routinely break the laws of physics. Take a simple example—in the movie *Spider-Man 2*, our hero stops a speeding train by standing on the front of it and spraying out spiderwebs, which attach to the buildings alongside the track, bringing the train to a halt. Leaving aside the biological discomfort that comes from realizing that Peter Parker shoots his webs from where spiders have their sex organs, not their spinnerets, the way this works is just wrong.

In the process, the spiderwebs are stretched for what seems like about half a mile—yet the fibers never get any thinner. Not only is there no material that could stretch that many times its original length, it would be impossible for the strands not to get thinner. Where is all the extra substance coming from? More worrying still, either the attachment points of the web to Peter Parker's wrists would break, the brick skin on the front of the building would peel off, or Parker himself would be torn in two by the amount of force being applied to him by the sheer momentum of many tons of speeding train. You could imagine a mechanism where the webs would work—repeatedly attaching between the train and buildings webs that stretch and snap, gradually slowing the train down over many miles—but the scene as played out is physically impossible.

Similarly, Superman's powers are a collection of nonstarters packaged up in blue tights and red shorts. When he flies up into the air, something has to be pushing back on either the air around him or the earth beneath using some kind of action at a distance—it's basic Newton's third law stuff—but what is it that does the pushing? Where is the force coming from? We are told he gets his powers from our yellow sun because his home world has a red sun. How? What physical forces are behind this ability? This doesn't even come close to being science fiction—it is pure fantasy. Or, if you prefer it, magic.

Now it might seem cruel to overanalyze our superheroes in this way. Notice the *f* word in science fiction—it's only a story. In the end, the science in science fiction is much less important than telling a good tale. And in most cases, worrying too much about the plausibility of the science is like breaking a butterfly on a wheel. Yet the point of making this unfair examination is that the restrictions and limitations we have to place on superheroes are due to issues with the physics. When movie directors want something dramatic to happen, they don't worry about what would physically be viable, they just do what looks good. In searching for real superpowers we don't have that luxury.

However, finding problems with superheroes' physical abilities leaves open the possibility of paranormal *mental* skills. Is there a loophole here that would make a form of superpower possible? The mind is much more mysterious than the straightforward laws of Newtonian physics. It's remarkable when you think about it—we know much more about how our galaxy, the Milky Way, works than we do about the functioning of the human brain. We learn more every year, but neuroscience has a long way to go to catch up with the other sciences.

It's hard not to wonder whether it would indeed be possible to go beyond normal human capabilities using the powers of the mind. There are some questions to ask, though, before we all don capes and tights. Do such paranormal mental capabilities even exist? Are there good physical explanations for [illegible] ents like telepathy and telekinesis? Or are the many repor[illegible] cases of such abilities just the result of wish fulfillment, ev[illegible] bit as much as classic superhero powers? Are all these examp[illegible] of supernormal capabilities nothing more than fraud or self-delusion?

It is true, as we will see, that we can conceive of scientific explanations for at least some of these so-called psi or ESP (extra sensory perception) abilities. Such mental powers could work within the bounds of physics, which surely makes them worthy of examination. But is there any real evidence to back up the existence of these talents, or are they nothing more than fantasy?

We should certainly not follow the lead of some skeptics and dismiss such powers out of hand. Science must always be open-minded (in this case, quite literally). To dismiss an observation without looking into it is totally unscientific. Admittedly a lot of effort has been put into the area in the past eighty years and it is hard to point to good, reproducible, incontrovertible evidence of the kind of psi abilities that we come across in anecdotes. But that doesn't excuse taking a stance based more on belief than on evidence.

All too often, scientists have a mindset that considers anything outside the currently accepted worldview as not being science at all. This is a sad mistake, because it demonstrates a deep misunderstanding of what science really is. As Professor Stuart Firestein points out in his book *Ignorance*, "working scientists don't get bogged down in the factual swamp because they don't care all that much for facts. It's not that they discount or ignore them, but rather that they don't see them as an end in themselves. They don't stop at the facts; they begin there, right beyond the facts, where the facts run out."

The point of Firestein's remark is not that we should ignore the facts and make science up as we go along, but rather that science is primarily about reaching out into the areas where we are ignorant. He likens the activity to searching for a black cat in a pitch-dark room, when the cat may not even be there. Although more aware of this aspect of science than the general public, scientists do still have a tendency to reject possibilities because they feel wrong or run counter to their personal beliefs, going along with the current majority view, even though this may mean missing out on vast swaths of juicy ignorance ripe for examination.

The late, great astrophysicist Fred Hoyle, writing about an alternative to the big bang theory in cosmology, used a picture of a flock of geese, all hurrying in one direction, to illustrate the tendency that scientists have to stick with the herd (or in this case, flock). The caption reads, "This [photograph] is our view of the conformist approach to the standard (hot big bang) cosmology." There's a balance, of course. It isn't possible to investigate every fringe idea and concept. Not all ignorance is equal in its value to science as potential fresh ground for exploration. But it is also true that science has to develop by taking a look at new possibilities, moving away from the flock of geese and taking a new viewpoint. We need to take the first steps into what is currently ignorance. The best scientists are able to leave their skepticism behind, even if they personally believe that something is highly unlikely, and really look at the evidence.

This is why it is still worthwhile examining the psi field even when many scientists say that there is no point, because there is nothing to examine. As sociologist Harry Collins points out, when a discipline is at the edge of understanding and the very existence of the phenomenon is doubtful, it is not at all uncommon for scientists to make outrageously ignorant statements along the lines of "There is no evidence whatsoever" or "Why investigate something that has no possibility of being real?"

As we will see, there is evidence of *something* behind claims of ESP—but the knee-jerk reaction is to manufacture criticisms based on personal prejudice. We need to take a step back and not prejudge.

The physicist John Bell, who made huge contributions to the field of quantum entanglement, an aspect of physics that some feel could describe a mechanism by which telepathy might work, put things very fairly when responding to a letter from a parapsychology researcher. Bell said that he was reluctant to criticize the problems of psi research. One big issue is repeatability. If scientific evidence is to be believed it should be repeatable in any appropriately equipped laboratory. This can be a problem with many aspects of psi research. But Bell had an interesting parallel.

Apparently, as a student in Northern Ireland, Bell was unable to reproduce standard experiments that demonstrate the expected response for electrical attraction and repulsion. These are basic physical experiments that have been accepted as the way things are since Michael Faraday's work in the nineteenth century. Bell said that he came to the opinion that "electrostatics could never have been convincingly discovered in my home country—because of the damp." Bell concluded that good scientists should certainly keep an open mind, as physicists had been surprised by seemingly impossible phenomena several times in the past.

Although there is no doubt that many who claim psi abilities are frauds, and no one has yet won James Randi's $1 million prize for demonstrating ESP under controlled conditions (see page 21), we still have a Nobel Prize winner suggesting a mechanism for telepathy, and serious scientists who have researched the field in major university projects like the Princeton Engineering Anomalies Research Laboratory who appear to have produced positive results of something above and beyond random chance. What's the verdict? Can we put any faith in the experiments that have been undertaken? That's the point of *Extra Sensory.*

Most books on psi veer to one extreme or the other. Some accept any "evidence," however flimsy the proof or however poor the controls. Others are based purely on attempting to show that everything in the field of study is bunk, the product of fraudulent practitioners and incompetent researchers. What I want to do is to go into this exploration open-mindedly, to examine the evidence, consider possible physical mechanisms, and come up with an educated view on the main fields of paranormal ability. I hope you are prepared to be equally open-minded, to decide on the evidence, not on your personal prejudices. There is a tendency sometimes to show that one particular aspect of the paranormal is totally worthless and to use that to dismiss all possible psi abilities. What we will do is take each potential ability and treat it in its own right.

Another problem that is often faced is defining just what is meant by psi, parapsychology, or the paranormal. All too often a whole mashup of psychic and paranormal concepts are bundled together as if they were in some way connected. The media have traditionally made few distinctions among all kinds of weird and wonderful phenomena. So you might see UFOs and abductees alongside psychic mediums who claimed to contact the dead alongside demonstrations of telepathy. I have endeavored to keep this book to topics that could have a physical explanation and that don't require a belief in spirits, fairies, or little green men. It's not that I am dismissing these possibilities (although I find many of them unlikely), but rather that I want to concentrate on the potential capabilities of the human brain—so not strictly paranormal, though definitely extra sensory—even though the claims for such mental powers are sometimes no less remarkable than the exploits of Spider-Man and Superman.

Some scientists are scornful, claiming that it's all over for paranormal abilities. They point out that traditionally many things that were once considered supernatural we now know to be either imaginary or the product of perfectly normal, natural phenomena. The supernatural aspects were first dismissed by science, and that dismissal has gradually been accepted by the general public. So, for instance, lightning was once seen as an unearthly force, quite possibly propelled by the wrath of the gods. Although there are technical aspects of the way that lightning is produced that we still don't understand, there are few people indeed who don't accept that lightning is a purely physical phenomenon, an electrical effect on a tremendous scale. It may be quite unlike the kind of thing that comes out of the socket at home, but it's electricity nonetheless.

If you look back at the remarkable summaries of thirteenth-century proto-science produced by natural philosopher and friar Roger Bacon in books like the *Opus Majus,* there are plenty of travelers' tales that we would now dismiss out of hand and wouldn't consider to be serious descriptions of the real world. You will find accounts of tribes of wild Amazon female warriors and mysterious devices for seeing at a distance that go beyond even the capabilities of a telescope. There are many marvelous, if unlikely, examples that were thought to be part of nature. For example, in his *Letter Concerning the Marvelous Power of Art and of Nature and Concerning the Nullity of Magic,* Bacon tells us

> that the Basilisk kills by sight alone, that the wolf makes a man hoarse if he sees him first, and that the hyena does not permit a dog to bark if he comes within its shadow. . . . Aristotle tells in the book *De vegetabilibus* that female palm trees mature ripe fruit through the odour of the males; and mares in certain countries are fertilized by the smell of horses.

These were all serious beliefs back then, as close to science as anything came. But these beliefs have joined leprechauns and fairies in the ranks of ideas that were not just incorrect observations of nature but totally fictitious. Just as these misinterpretations and fictions have disappeared from everyday life, it is argued by some skeptics who can't even be bothered to examine the evidence that telepathy, remote viewing, telekinesis, and the like have also reached a stage where they should no longer be considered anything more than a fantasy or a historical misunderstanding.

I would suggest that we have not reached that stage while there are so many people who still think that there is something to be investigated, and while a host of experiments have thrown up evidence that at least needs careful examination. We ought to take a look at that evidence with fresh eyes, biased neither by enthusiasm for the topic nor by scientific blindness that refuses to even look at the evidence because we "know" there is nothing to see.

Let's make our first steps into the unknown and take on ignorance.

2.

PARAPSYCHOLOGY—SEPARATING SHEEP AND GOATS

In the eerie green glow produced by a night-vision camera we can make out four people sitting around a table. They are holding a séance in a location that is infamous for its frequent ghostly activity. It is difficult to detect emotion in the scarily blank eyes that the camera gives to the participants, who are being filmed for the British TV show *Most Haunted.* Are they terrified or just amused by it all? The show's host, a onetime children's TV presenter called Yvette Fielding, clearly takes her role very seriously. She is quoted as saying "There is no acting in this programme, none whatsoever. Everything you see and you hear is real. It's not made up, it's not acted."

On the screen, the table begins to rock from side to side, vibrating with an uncanny regularity. It is as if it moves with the pulse of something living. Some unseen force is driving it. The large candlestick in the center of the table vibrates as the tension rises in the voices of those taking part. For most viewers that was the end of the experience. A frisson of excitement, a touch of entertainment and amazement at the wonders of the psychic world, and then on to another show. A few, though, noticed something strange. They rewound and watched again to make sure they hadn't been mistaken. Yes, there it was. The tablecloth was rippling rhythmically under Fielding's fingers. Exactly as if she was pushing the table to produce the effect, and the pressure of her fingers was crumpling the cloth.

Whether or not there was cheating, this is exactly the kind of phenomenon I want to avoid in this book. The best blanket term for the mental abilities we will cover, abilities that go beyond the everyday (and still remarkable) capabilities of the brain, is probably "parapsychology." Though this isn't fully recognized as a scientific discipline, the intention is to cover only those abilities that could be due to as yet unexplained functions of the brain, while excluding possibilities that are either pseudoscience or make claims that are not accessible to scientific testing. Fielding's ghost hunt claimed to be detecting the actions of an independent spirit rather than an extension of the mental capabilities of those present, which puts it beyond our scope. And what was being demonstrated was certainly not being properly tested.

It may seem very arbitrary to dismiss something just because it can't be tested and measured. Many New Agers would get worked up at such a move. They would argue that it demonstrates the closed minds and shortsightedness of the scientific community, unwilling to work with anything that doesn't fit with their cozy picture of reality. But sticking to studying paranormal activities that can be tested, repeated, and quantified is essential if we are to make any sense of claims for special abilities. This limitation does not say that something doesn't exist if it can't be subjected to scientific scrutiny, just that science can have nothing useful to contribute on the subject, so there is no point in trying.

This is what is sometimes called the "invisible dragon" phenomenon. If I tell you I have an invisible dragon in my garage, I can make it possible to totally exclude any scientific testing of this claim. For example, if you try to detect the presence of my dragon by scattering flour on the floor to pick up footprints, I will point out that my invisible dragon is weightless, floating just above the surface of the ground. If you listen for its breathing I will tell you that my dragon does not breathe. If you want to detect it with a heat gun, I will point out that it emits no radiation. My dragon *could* exist. But if there is nothing to measure, nothing to observe, science can say nothing useful about it.

There are even some areas of science itself—string theory comes to mind—that probably should fall into the invisible dragon category. Hundreds of physicists are working on string theory, yet at the moment it makes no testable predictions that could be used to prove that it is true. The search for a unifying theory for gravity and the other forces of nature, such as string theory, has been described as the hunt for a "theory of everything." As physicist Martin Bojowald has pointed out, string theory as it stands is a literal theory of everything because everything—and anything—can happen within it.

As such, arguably string theory is not science. And the same probably applies to ghost hunting. Other paranormal claims like the existence of reincarnation, the abilities of spirit mediums to contact the dead, or the power to receive electronic messages from the afterlife also rely on assumptions that go well beyond the physical world. What we need to do here is to identify the capabilities that could have physical explanations that fit within the ability of science to produce explanations and that could be proved or disproved by experiment and testing. And specifically, for the purposes of this book, we need to focus on abilities that are linked to the human mind.

In doing this I am not dismissing the possibility that the claims of spirit mediums, for example, have some validity, although it is tempting to do so. The idea of especially talented people being able to contact the dead has a long tradition, stretching back into ancient times. By the nineteenth century mediums had settled down into the broad approach you can still see practiced in theaters and spiritualist venues around the world. Yet most of the early spirit mediums produced effects that today would be considered laughable—while they worked in complete darkness, luminous trumpets would be lifted into the air and blown, curtains would be felt to billow out, a strange glowing substance given the name ectoplasm would flow from the medium.

Time and again these individuals were revealed to be frauds. Some of the best known were caught performing tricks like clicking their toes to create spirit "raps," or using an impressively lithe but in no sense psychic twist of their leg to set a curtain in motion. If all else failed they would resort to having an assistant in the darkened room perform their miracles for them.

From a modern-day perspective it is difficult to understand how our predecessors, often educated and sometimes scientifically trained people, were taken in by these charlatans. Even in complete darkness it is hard to believe that anyone would think that a piece of muslin soaked in luminous paint was "ectoplasm" or that the dead would be so enthusiastic about lifting and blowing luminous trumpets. Apart from anything else, it seems highly unlikely that a modern audience would be able to resist making a grab for such props (something that did occasionally happen even back in the day), proving the lack of psychic force behind them.

In part because of the sterling work of the likes of Harry Houdini in unmasking the fake mediums, there was a significant drop in the claims made for these specialist psychic workers. One kind of psychic mediumship has continued unabated, though. This is the stage performance of a mental medium. Here, rather than working with physical objects, the medium attempts to bring messages from the dead to audience members. Superficially impressive, these performances (and that's what they are, with a paying audience in attendance) are master classes in "cold reading" where the performer picks up information from audience members without those present realizing. This is often combined with a spot of research, making careful observations of audience members before the show.

Once you are aware of the technique that is used, it becomes embarrassingly easy to spot the cold reader in action. Imagine that a medium really did have a message from the dead—it would likely be along the lines of "Betty has a message for her son Bill. She is sorry about the silly section of her will leaving most of her money to the cat, but she has left a small cache of cash behind the vase in the lounge—the two-foot-high green one she used to put sunflowers in. Oh, and she really wishes her great-grandson hadn't been called Moonstone." Such a specific, easily checked message would be impressive indeed, and would suggest that there was a need for further investigation.

Instead, an expert in cold reading will deliver something like "I've got someone whose name begins with B. . . . She's passed away fairly recently, and her son or daughter is here?" Long pause. Bill pipes up: "My mother, Betty, just passed over." "That's her," responds the medium. "Your mother is here. She misses you. And she's sorry if things weren't always right between you. She realizes it was mostly her fault." Wow, thinks Bill. This medium is great—she knows about that problem with the will. And so it goes on, with the medium uttering vague pronouncements and generalities that could apply widely and that can never really be checked for accuracy.

According to the mediums, these messages are so vague because the spirits are detached from the real world, or not entirely concentrating, or . . . whatever excuse they can dream up. But in reality they are engaged in a fishing exercise, latching on to anything they throw out that gets a positive response. What is fascinating is that if you watch a cold-reading medium at work, he will often strike out. He will say something that gets a blank response and quickly move on. But afterward, the happy recipients of readings will have little recall of where the guesses went wrong. They won't wonder why their departed mother had got the idea that they had a sister when they never did. (Some mediums will even twist this, suggesting a sister that the mark didn't know about, or someone who was "like a sister to them.") The audience will just remember the good bits.

Although displays of physical mediumship—the séances featuring glowing trumpets and ectoplasm—had largely died out by the 1970s, there has been something of a revival as a result of the so-called Scole experiment, named after the village of Scole in Norfolk, England. From 1993 to 1998 a group in Scole undertook a large number of physical mediumship events, claiming a whole host of effects that were reminiscent of the classic séances, from mysterious floating lights to those present being touched by an unseen hand. There were also some modern twists, including "psychic photography," where images appeared on photographic negatives contained in sealed boxes during the sessions.

At first sight some of the activities at Scole seem very impressive—but when you come down to detail, these spirits seemed to have the same shyness as their forebears did for anything that could show them up as fakes. If darkness is truly required for a séance (though there is no good reason for this, apart from making it easier to cheat), then we have cameras that work happily in the dark now using invisible infrared. But no, these could not be used at Scole. Apparently these would upset the modern-day spirits.

Whenever the independent observers suggested some form of control, the mediums involved would say that they were not possible. When an observer occasionally got control, there was no psychic experience. For example, when observers substituted their own box for the mediums' container in which the film to be exposed was contained, no images were produced. Although Scole is often presented as the most scientific attempt ever to study physical mediumship, in practice science was not allowed to have any say in the proceedings. The only real controls were luminous wristbands, supposedly impossible to remove without being detected (a claim that was never tested). These amounted to no control at all. Take apart the claims for Scole piece by piece and there were plenty of opportunities for cheating and misleading at every turn.

This is very much the kind of thing that may spring to mind if we cast a wide net and look at all psychic phenomena. But the evidence against mediums of all kinds is strong, and even if we take them at face value, they are not claiming the kind of mental powers that science can investigate. Instead, I would like to consider those areas of mental ability—extra sensory perception and telekinesis, for example—that could still be real, could have an explicable scientific basis, and have been demonstrated in at least partially controlled experiments in laboratories.

A major problem faces scientists who study psi phenomenon. They have to be constantly aware of an issue that most of them will not have encountered in their everyday work. Take, for example, physicists who become involved with investigating psi claims, which many have done because we would expect some sort of physical phenomena to lie behind the ability. The physicists may well be very experienced experimenters with years of lab work behind them. But the physics experiments they are familiar with don't cheat. Electrons don't decide suddenly to behave as if they have a positive charge just to fool the experimenter. People, however, do cheat. And this possibility has to be allowed for in the experiments.

It might seem very mean-spirited to assume that participants will try to mislead—but we know from experience that they have often done so. When making claims that would mean a major breakthrough in science, the controls to avoid cheating have to be strong. When those who have been caught out are asked why they resorted to cheating, they often admit to a sense of mischief, an attempt to get publicity, or giving in to the urge to fool academics who are supposed to be cleverer than they are.

I know the appeal of cheating to produce strange phenomena—when I was in my teens I did quite a lot of fake UFO photography. I never tried to sell it, it was just for fun—but it was enough to make me realize that it is very naive to suggest there is no motive for people to cheat. In practice there are many reasons this can happen. The deception could even be coming from the experimenter who is determined to make the point (or keep his funding). This may be conscious or subconscious—it is entirely possible, as we will discover later, for scientists to see the results they want to see in an experiment where the outcome is subject to interpretation. But the majority of the cheating seems to originate from the experimental subjects.

Often, I suspect, this is simply the result of an urge to please the experimenters. The subjects know that the researchers want to find evidence of psi abilities—the subjects may even be financially rewarded if they can demonstrate them—which means they are prepared to go out of their way to provide that evidence. But this does not mean that all attempts at pulling the wool over scientists' eyes are about being helpful. There is entertainment to be had from fooling experimenters who are apparently smart and perhaps more than a little pompous. It is instructive to take a quick look at one of the earliest examples of a scientifically controlled demonstration of telepathy that later proved to be fraudulent at the admission of one of the subjects.

To examine this we are going all the way back to the 1880s, when two individuals, Douglas Blackburn and G. A. Smith, gave dramatic telepathy demonstrations that were accepted at the time as genuine by the British Society for Psychical Research, the main body attempting to apply scientific methods to understanding psi phenomena back then. In a typical session, Smith would be placed in a chair and controlled by many mechanisms that were supposed to prevent sensory contact with the outside world. He was blindfolded, his ears were stopped, and he was swathed in heavy blankets, all in an attempt to prevent him from communicating with his confederate—though the blankets would prove to be a serious mistake on the part of the experimenters.

Blackburn would attempt to send words and images to Smith by thought alone. And surprisingly frequently Smith would manage to reproduce the information within the depths of his blanket prison. It all seemed aboveboard, but twenty years later Blackburn came clean. He admitted that "with mingled feelings of regret and satisfaction" he was revealing that all the experiments were bogus. There was no telepathy occurring in the experiment. With surprising insight into the human psyche, Blackburn said that the trickery originated in an "honest desire of two youths to show how easily men of scientific mind and training could be deceived when seeking for evidence in support of a theory they were wishful to establish."

This urge to make the investigators (often, dare we say, full of their own self-importance) look foolish is likely to be a common thread in many of the ways that amateur subjects have misbehaved during psi research. (I use the term "amateur" here to distinguish them from those professionals who have made a living out of claiming to have mental powers and who have rather more obvious financial reasons to deceive their audiences.)

Blackburn went on to explain how he and his partner had cheated in one session, describing a remarkably sophisticated method to get around the controls, considering that neither of the young men was a stage magician. When Smith was swathed in blankets, Blackburn secretly drew on a piece of cigarette paper a copy of the image to be mentally transferred. He then concealed this miniature document inside a mechanical pencil. When it was in place, he signaled to Smith that he was ready by stumbling against the edge of the thick rug on which Smith's chair stood.

On feeling the vibration, Smith shouted out that he was picking up a message, and felt around on the table in front of him, saying, "Where's my pencil?" By this time, Blackburn had casually dropped his pencil on the table, and it was this that Smith picked up. Underneath his blanket, Smith had a luminous slate, which he used to read the message on the cigarette paper, dislodging his blindfold just enough to be able to peer along the side of his nose. (This is a standard magician's trick, as practically all blindfolds can be got around this way.) Smith was then able to duplicate the image under his blanket, which may have been designed to prevent communication from Blackburn, but in practice gave Smith ample opportunity to cheat.

Those witnessing these demonstrations were not stupid. They were trained scientists who thought that they had established adequate controls to ensure that no communication was taking place. They were far better educated than Blackburn and Smith—in fact, these scientists were men of the world, so they thought. Yet, as Blackburn put it, "two youths, with a week's preparation, could deceive trained and careful observers . . . under the most stringent conditions their ingenuity could devise."

The revelations in the press of these and other early deceptions should have made for much greater care in designing controls in future psi experiments—but as we will see, this was often not the case. Frequently those "trained and careful observers" were deceived by simple means of misdirection and trickery. Not in all psi experiments, certainly. But the ghosts of Blackburn and Smith have continued to haunt the scientific investigation of the psi world ever since.

Time and again, going all the way back to the early-nineteenth-century investigations of psychic phenomena, scientists have been caught out by cheats using unsophisticated methods (see chapter 10 for some examples of gullible scientists). And this is where a name that will crop up regularly through this book comes in. Over the years one man has stood out from the crowd in his stance against fraudulent claims of ESP and other psi abilities: James Randi, aka the Great Randi. This retired stage illusionist has specialized in uncovering and challenging mental abilities.

In the early days of scientific investigation, it seems, leading British physicist Oliver Lodge, who was a strong supporter of psychics, fell for a whole range of fakery. More recently we have seen academics from universities in both the United States and the UK bamboozled by slick professionals. But it is much harder to pull the wool over the eyes of magicians who can fake their own psychic phenomena to order, as Randi and his British equivalent Derren Brown have shown.

Randi set the bar high relatively early on by offering a reward for anyone who could produce psi effects or other paranormal abilities under properly controlled conditions. Originally the sum offered was $1,000, but it has since grown, and his James Randi Educational Foundation now offers a $1 million prize to anyone who can demonstrate psychic or paranormal abilities "under satisfactory observation." It is this condition that cuts through and disposes of so many claims for psi abilities. Without a combination of the strictures of science, and a stage illusionist's ability to spot a fake, there is little hope of ever having a true scientific picture of such capabilities.

As is often the case with a strong personality opposing a particular belief (think Richard Dawkins taking on religion), Randi can come across as both self-promoting and smug. He has a tendency to make blanket statements like "There is not a single example of a scientific discovery in the field of parapsychology that has been independently replicated," which is not strictly true. But he has done an immense amount to uncover fraudulent practice in ESP and has usefully pointed out the limitations of many apparently scientific explorations of the subject. It is notable that many well-known psychic performers, like Uri Geller, refuse to perform if they know Randi is present. They will no doubt say that his malign vibrations suppress their abilities, but in reality there is little doubt that they are terrified that he will uncover their methods.

The "Million Dollar Paranormal Challenge" run by Randi's foundation is not really relevant to the scientific investigations of psi that form the majority of the content of this book, but it is hugely important as a means of weeding out the flamboyant claims of psi abilities made by some individuals. It's difficult to ask why, if they are genuine, they haven't applied for and won the million dollars. Admittedly the process is a little drawn out, and applicants have to cover all costs, which without an indication of what these costs might be is more than a little worrying. But anyone knowing he had genuine abilities would take those costs on board to get his hands on the million. The organization gets about one serious application a week, and as yet no one has come close to winning. The process has a preliminary test before the formal testing under fully controlled conditions. At the time of writing in 2012, not one single application has got through to the formal test.

There are limits to this million-dollar challenge. As Randi's organization comments, it won't test claims that could cause injury: "For example, if you claim you can jump off a ten-story building and survive, the JREF is not going to test you at it because people jumping off buildings doesn't normally end well." And the approach that the foundation uses is really designed more to deal with the dramatic personal claims of showmen than with the subtle statistical effects that some laboratories claim to have discovered. But it was a brave challenge to make—originally making use of Randi's own money—and is a useful filter for charlatans.

The fact is that if laboratories *are* to test individuals who make extraordinary claims, they would be very wise to call in assistance from someone with the experience and expertise of James Randi in setting up controls that would make it nearly impossible to cheat. Even psychologists, who generally have a much better grasp than physicists of just how and why their subjects are likely to try to cheat, need that assistance because they aren't aware of the techniques available to give the impression of performing paranormal feats based on magical tricks and misdirection.

So if we are to pick out the real thing, what are we actually going to look at in terms of "extra sensory perception" or "psi abilities"? It's worth outlining first what our "sensory perception" is, as we can hardly consider what falls outside that without first being aware of the remarkable capabilities of our more familiar senses. Our senses are key to our interaction with the world around us, the gateway between our brains and the universe, yet most of us are surprisingly vague about them.

Everyone will be familiar with the idea that we have five regular senses; it's something we are taught as a matter of course. Ask anyone how many senses there are and she will come up with the number five. (Even the expression "sixth sense" is dependent on there being five everyday ones.) Yet the idea of there being five senses is total baloncy. This is not denying the existence of sight, hearing, touch, taste, and smell, but simply pointing out that there are many more senses we make use of every day. Imagine you were suspended upside down with your eyes closed. Leaving aside touch, feeling the pressure on your legs at the suspension point, how would you know you were upside down?

Similarly, imagine that you had your eyes closed and someone brought a red-hot poker toward your bare arm. At what point would you be aware of the poker? And how would you know it was there? If you relied only on the traditional five senses you wouldn't know the poker was there until it actually touched your arm—yet this clearly isn't the case in practice.

Being upside down, you would feel the blood rush to your head, but none of your basic five senses are responsible for this sensation. If you happened to be upside down because you were on a roller coaster you would experience all the usual sensory inputs, but even if you took them away you would be aware of the acceleration as you powered around a curve or down a drop. This is because, just like a smartphone, you have a built-in accelerometer. In the case of a human being, this accelerometer consists of fluid moving around in the inner ear to detect rotations and crystals sliding on a gelatinous goo that moves under linear acceleration. These movements disturb hairlike sensors. They are primarily there to help with your balance, but they do so by detecting acceleration.

Then think of that red-hot iron coming close to your arm. The first thing you would detect is heat, well before the poker came into contact with your skin. This is not touch, but the detection of heat by your skin. In effect, in a very crude unfocused fashion, your skin can "see" infrared light. You can feel the incoming infrared as a warm sensation. If the poker came close enough you would also feel pain. This is quite different from touch—in fact, you can have a pain from burning without anything touching you at all—though pressure-related pain goes alongside the tactile feeling of something pressing into your skin.

Pain can be produced by plenty of the standard sense detectors—too bright a light or too loud a sound, the burning pain of chili powder on the tongue or the assault on your nose of a violently pungent odor—but it is a totally separate sensory experience from those senses. If you doubt this, try getting a small amount of chili powder on your nostrils or at the edge of your eyes, as many people manage to do while cooking. You will still feel an unpleasant burning sensation, but you would hardly say that you are tasting it with your nostrils or eyes.

For another example of a sense that isn't one of the big five, try closing your eyes and touching your nose. The chance of hitting it randomly without a sense to direct your hand is pretty low—but the fact is that you should be able to do it every time. You clearly aren't using any of the conventional senses. This ability is attributable to a sense known as proprioception, which is the awareness of the location of parts of your body. You don't have to see or feel anything to know where your head or hands are—feedback from your muscles combined with an awareness of your body scale provide the appropriate information.

Our extra senses on top of the famous five go only so far, but some animals have senses that we can only imagine. These might extend a sense we do have but take it far beyond our capabilities. Dogs' sense of smell, for example, is capable of detecting odors a million times weaker than anything we can pick up, and they seem to build a three-dimensional mental smellscape as they make their way through the world. Hawks have an extra color receptor in the eye on top of our red, green, and blue that enables them to detect ultraviolet, useful in finding their prey, for mice constantly urinate, leaving a trail that is highly visible in the ultraviolet region. And, of course bats take hearing to a whole new level with their echolocation skills.

In fact, to call echolocation "hearing" misunderstands what is going on. It is a totally different sense. It happens to use the same organs as hearing—but it is closer to vision in the kind of information it provides to a bat's brain as the animal flits about in darkness, effortlessly missing obstacles. Other creatures also have senses that we can really parallel only with technology. Sharks are a great example with their apparently uncanny skill of detecting prey in murky waters despite having bad eyesight. Many kinds of sharks have the ability to sense electrical fields, even the weak electrical fields generated by the nervous systems of their prey. Electromagnetism also manages to get into the heads of some birds, which are able to navigate using a kind of inbuilt compass that detects the Earth's magnetic fields and guides their migration.

So when we say that in parapsychology we are looking for abilities that are extra sensory, we may just be looking beyond the traditional five senses, or even outside those extra animal abilities, to something as yet unknown, though the animal supersenses may give us some small hints of ways that new human senses and means of communication could function.

It's time, then, to begin our exploration, with what is probably the most widely demonstrated and certainly the most widely believed in of the psi abilities: telepathy. A term devised back in 1882 by British psychic researcher Frederick Myers, it literally means "distance feeling," but the ability it describes is closer to mind reading—mind-to-mind communication without using the usual senses. Put aside that early example of cheating in mental communication: we are going to see what is really going on.

3.

CAN YOU HEAR ME?

We are on the island of Colonsay, part of the group of small islands that form Scotland's Inner Hebrides. It is 1967, but from the pace of life and the things we see around us, it could just as easily be the 1940s. Things move slowly on this island, warmed by the Gulf Stream so much that semitropical plants grow in what should be chilly northern climes. There is still a weekly ceilidh, the Gaelic equivalent of a barn dance, held in the island's single hall. Practically every inhabitant goes to one of the island's two fiercely plain Protestant stone churches on Sunday. You only have to walk for five minutes along the island's single paved road and someone will stop to offer you a lift, a lift that you will happily accept. It is like passing through a time warp.

If you are lucky, the person who stops to pick you up might be one of the island's great eccentrics, the doctor. This graying, middle-aged man drives an open-topped Land Rover in all weathers. Beside his driver's seat is a shotgun. You had better be prepared to hang on for dear life—if he spots a rabbit on the moor, he will be off-road in a moment, jolting across the open fields after his prey, driving with one hand and holding the shotgun with the other. But he is courteous enough. He won't be offended if you would rather not get out of the car and pick up the dead animal for him.

There is a camp of schoolboys on one of the island's startlingly white beaches, the silvery sand giving the waters of the bay the azure beauty of a Caribbean paradise. One evening as the sun sets, a lone piper passes by. There is no explanation for his presence, just the skirl of the pipes coming over the sand dunes, closer and closer. He walks along the ridge and disappears again. Even those who have never liked the sound of the bagpipes can understand the sheer gut impact of their emotional appeal in this setting.

The morning after the piper's visit, there is something of an emergency. One of the boys needs to get his best friend back up to the camp, but the friend is running down the beach, following some unknown errand. The second boy tries to catch him, but his friend is too fast. There is nowhere near enough breath in the pursuer's body to be able to shout. In his head, he is calling out, "Paul, stop!" but he can't speak; all he can do is pound on across the sand. His friend is at least a hundred yards ahead, and a stitch suggests the chase will soon be over. Suddenly the friend stops and turns, heading back toward his pursuer. When asked why he suddenly stopped, he is puzzled.

"You shouted," he says. "I heard you shout, 'Stop, Paul!'" But no one made a sound. That cry had been purely mental.

This happened to me at the age of twelve, the only time I have ever directly experienced what could be a psi phenomenon. I was the pursuer. Even now, over forty years later, I have a clear memory of this happening. I gave so much detail because context is very important when analyzing a parapsychology event, especially an anecdote like this. It is all too easy to remember something dramatic—perhaps embellished a little by false memory—and to present it in isolation. This is a mistake often made both by parapsychology researchers and by those who are taken in by frauds and entertainers.

The researchers, for good scientific reasons, try as much as possible to isolate the occurrence they are studying from a normal situation. An essential for good scientific research is control. But if there is any truth in most accounts of naturally occurring psi phenomena, the way most research has been undertaken seems designed to prevent anything from happening. Telepathy, for example, is consistently reported to occur between those who are close to each other or have a strong connection in some way, and also to happen most frequently under pressure or where there is a particularly urgent need to get a message across. Yet almost all the laboratory trials of telepathy have been between strangers, with no urgent need for communication whatsoever. Some experiments even involve sensory deprivation, minimizing any feeling for place and urgency.

It's as if you wanted to study earth tremors and decided to do so in a location with a very stable geology, in a building that had shock absorbers. This is surely a mistake. It is perfectly possible (if rather harder) to apply controls with situations that are more conducive to telepathic communication, should the ability exist. At the extreme, taking away the essentials is a bit like doing a medical test to see how a drug helps patients with diabetes, but only testing the new medicine on people who don't suffer from the disease.

Removing an occurrence from its environment, reporting

it in isolation, is also a problem where a gullible person is being fooled by someone using trickery to fake the appearance of a psi ability, either for profit or for entertainment. If you simply describe the immediate event without any context, it can be very misleading. In a famously flawed study of Israeli mentalist Uri Geller (of whom more in chapter 10), it was reported that Geller repeatedly predicted the results of a dice throw that took place inside a closed box. What wasn't reported but subsequently came out was that Geller was able to handle the box before the results were produced. That element of context makes all the difference to the importance of what was described.

If we are to examine my own experience, there are a number of contextual items to be considered. On the positive side, I was trying to communicate with my best friend, and I was under pressure—mentally I was screaming inside my head. On the negative side, we were in an environment that was already unreal-feeling, perhaps making it easier than usual to construct artificial memories. How much of my recall from forty years ago is truly accurate has to be open to question. Most important, even though I was sure at the time that I was too out of breath to produce a sound, and much too far away to be heard, could I have called out and not realized it? Could this be simple, everyday sensory perception of the human voice?

Telepathy seems particularly natural because communication between human beings plays such an immense role in our lives. Ever since writing was developed we have been able to put across ideas and information at a remote location, as if we were present. In recent years we have seen the opportunities for this distant communication explode with the advent of the telegraph, then telephones, radio, and the Internet. I now take it for granted that I can communicate pretty well instantly with someone on the other side of the world via e-mail or social networking, or verbally and visually with a product like Skype. Given the everyday aspect of person-to-person communication, ranging from direct speech to these technological solutions, telepathy does not seem so strange.

What's more, of all the psi abilities, telepathy is arguably the easiest to fit in with a scientific viewpoint of the world. When considering the possibility of a parapsychology event's being genuine we have to look at the basic physics. How much energy was required? Where did that energy come from? How could information be transferred from place to place? How many complications were involved, like being able to see through time or being able to manipulate objects remotely? Telepathy scores well across the board. In the end, it is simply a matter of communication, something that science enables us to do all the time. More people claim to have experienced telepathy (myself included) than any other of these phenomena, making it a good starting point for investigation.

What would be useful in understanding the research to date is to establish whether there is any possible mechanism for telepathy to use to send a communication directly from brain to brain. In 2001, Nobel Prize–winning physicist Brian Josephson caused a furor when he suggested that telepathy could be a product of quantum entanglement, famed as one of the strangest phenomena in all of science ever since Einstein first referred to it as "spooky action at a distance."

Josephson is, without doubt, a very intelligent scientist. He won the Nobel Prize in 1973 for his theoretical predictions of the ability of a special kind of quantum electrical junction, now known as a Josephson junction, to allow superconducting effects to take place across a barrier. This is high-powered physics indeed. But Josephson also displays far more open-mindedness than most scientists. (His opponents would say that he is altogether too open-minded.) He is prepared at least to consider far-out ideas like the memory of water, used by some to try to explain the effects of homeopathy, that most dismiss out of hand. And some would say that he exhibits a naïveté when presented with information and demonstrations that are easily dismissed, a fault that has always dogged scientists, particularly physical scientists who have attempted to investigate phenomena outside of the mainstream of science.

Josephson's venture into the theory behind telepathy came up, strangely, as a result of a special issue of stamps. The UK's Post Office produced a set of commemorative stamps on October 2, 2001, to mark the one hundredth anniversary of the Nobel Prize. To encourage collectors to buy them, stamps like this are available in a presentation pack, which usually contains some written material—a bit like the bonus features on a DVD. In this case it included a "Nobel reflection" by each of six British prizewinners, one for each of physics, chemistry, medicine, economics, peace, and literature. Josephson contributed the physics entry.

The contributors were given free rein as to their subject, and Josephson, never one to avoid controversy, decided to stir things up—when later asked if his intention was provocative or serious he responded, "Both." In the short article he wrote, Josephson introduced quantum theory and stressed its theoretical and practical effectiveness. He finished by saying that it was now being combined with computing and information theory and that these developments might lead to an understanding of processes not yet explained by conventional science, such as telepathy.

The storm in response to this publication was not long in coming. Physicists were quoted as saying everything from the mild "I am very uneasy" to the dramatic "It is utter rubbish. . . . The Royal Mail has allowed itself to be hoodwinked into supporting ideas that are complete nonsense," which came from leading Oxford scientist and longtime Josephson opponent David Deutsch. The Royal Mail's spokesperson Kathryn Hollingsworth weighed in with "If it transpires that what he's suggesting doesn't have a scientific basis, perhaps we should have checked that, but if he has won a Nobel Prize for his work, that should give him some credibility."

Josephson was given a chance to respond to his critics on the flagship BBC radio current affairs show, *Today*. He was introduced with a recorded clip of psychic debunker James Randi in which Randi, with typical forcefulness, commented that "there is no firm evidence for the existence of telepathy, ESP or whatever we wish to call it, and I think it is the refuge of scoundrels in many aspects for them to turn to something like quantum physics, which uses a totally different language from the regular English that we are accustomed to using from day to day, to merely say, oh that's where the answer lies, because that's all very fuzzy anyway."

When faced with this, Josephson suggested it was worth considering the work of Henry Stapp of the University of California, who has suggested that science has not taken the mind properly into account. In essence, Josephson said, Stapp is reflecting on the long-established aspect of quantum mechanics that the observer has to be taken into account—that he is part of the system. "It's not a crazy idea," said Josephson; "it is absolutely standard physics." As we will see, this assertion slightly stretched the truth but was certainly not pure fiction either. Arguably both Josephson and Randi were wrong.

The main arguments used to counter Josephson's article were based on the dubious logic that telepathy doesn't exist (with no clear basis for this assertion), so there is no need to establish a rationale for it. We will come back to the evidence for the existence of telepathy, but is quantum theory really able to provide a mechanism for telepathy to work—or is this just a case of fuzzy thinking and confusing jargon, as Randi suggested? There are two parts to Josephson's comments: to suggest that quantum entanglement might be involved and to point out the essential role of the observer in quantum physics. Both of these need a little explanation.

One essential component, if quantum physics is to describe for us a vehicle by which telepathy or other psi phenomena work, is to somehow link a human being's brain to an external physical result. The sheer effort of thinking at something has to produce an action somewhere unconnected to your body if we are to communicate remotely. Quantum physics does indeed provide just such a mechanism based on the importance of an observer in quantum systems—possibly. There are provisos attached.

A quantum particle like a photon of light or an electron in a metal wire is quite unlike the macro equivalent of such a particle—a tennis ball, say. Imagine I put a tennis ball on a table and start it spinning with a flick of the wrist. I can now look at that tennis ball and tell you how it is spinning (it might, for instance, be spinning clockwise as seen from the top). What's more, my looking at it will have no effect on the way it spins. There is no significant interaction between my making the observation of the ball and the spin itself.

Now let's take a quantum particle. Just like my tennis ball, it has a property that physicists call spin. They gave it this name because there are some similarities between quantum spin and real spin. For example, quantum spin undergoes a kind of conservation of angular momentum, the idea being that the rotational oomph of a body stays constant unless a force is applied. However, quantum spin is actually nothing like the spinning of a tennis ball. We have no reason to think that a photon is actually spinning around; the term "spin" is just a label. We could just as easily have called it quantum personality, or quantum politics. And this becomes clear when we try to measure that property.

If I take a photon and measure the spin, it can have only one of two values, up or down, corresponding to the two possible directions of rotation in real spin. But here's the interesting thing. Before I make the measurement, the particle is in what's called a superposition of states. It is both spin up and spin down at the same time. (This is where the idea of the infamous Schrödinger's cat, which is both alive and dead, comes from. More on that in a moment.) The particle might have, say, a 60 percent chance of being spin up and a 40 percent chance of being spin down. If I measure the spin of particle after particle, 60 in every 100 will be up and 40 will be down, but before making the measurement, there is no way of telling what the outcome will be for any particular particle.

This is the bit of quantum theory that Einstein hated, the aspect that he fought against for decades and that made him complain that God does not throw dice. It's not that before we make the measurement, 60 out of those 100 particles are in a spin up state and 40 are in a spin down state. Before we check them, they are *all* in a superposition of up and down, with a 60 percent chance of being up after measurement and a 40 percent chance of being down. If it helps, you can think of the spin as being a direction between up and down at an angle with a 60:40

ratio between the up and down directions. There is no secret information that tells a particular particle which state it is "really" in. It is genuinely in both.

Here's the bit that is of interest to psi researchers. All it takes to switch a particle from being in the superposed state, both up and down simultaneously, to a single "real" state—a process known as collapsing the waveform—is to look at it. This "waveform" is the outcome of Schrödinger's equation, the central plank of quantum theory, which describes the state of the particle—or a collection of particles—as a probability wave, a state that varies in space and time according to the equation, which describes how this waveform is structured.

When you make a measurement, the particle is forced into one of the two states. There is a lot of argument over just what "looking at it" entails. But if "looking at it" means a conscious mind accessing the information, something some physicists following the thinking of Hungarian American physicist Eugene Wigner believe to be true, then what we have is a direct link between the conscious mind and a quantum event that is remote from it—a potential building block for a mechanism that could allow telepathy to work.

It's worth spending a minute with Schrödinger's cat, despite how frequently this thought experiment is described, to see the pros and cons of the idea that a conscious mind is required to collapse the waveform. In the cat experiment (never carried out, incidentally—no cats have been harmed in the making of this theory), a cat is put in a box with a radioactive particle that we know will decay at some point in the future. Like the spin of the particle, the timing of the decay does not have a fixed value but has various probabilities attached to it. This is why we measure the lifespan of radioactivity in terms of half-lives. We can't say when any particular particle in a lump of radioactive material is going to decay. At all. But we *can* say how long we have to wait before half the particles (we have no idea which) will have decayed—the half-life.

After a period of time has elapsed, the particle inside the cat's box is in a superposed state. There is a certain probability for its having decayed in that time period and another probability for its not having decayed. But it is impossible to say what the state is without looking, because the particle is in a superposition—it is genuinely in both states at once. As well as the cat, there is a detector inside the box that will be triggered when the particle has decayed and a device linked to that detector that releases deadly gas into the box. So if we discover that the particle has decayed we also know that the cat is dead.

However, think of what things are like inside the box before we take a look. As the particle has not been observed, it is in both states at once—which means the device both has been triggered and hasn't. Not "it may or may not have been triggered"—both states exist simultaneously. Which means the cat is both alive and dead at the same time until we look in the box, collapse the waveform, and produce a definitive result, one way or the other.

Many people would argue that deploying the detector constitutes making an observation, so in practice we don't have to open the box, because the particle is constantly being observed. But others say that what we have to consider is the quantum state of the whole setup. After all, the detector is made up of quantum particles itself with their own contributions to the waveform that describes the quantum reality of the whole—in which case there really is an alive-and-dead cat before the box is opened.

This is the effect that those who believe that there is a quantum mechanism behind telepathy build on. A human consciousness appears to have an impact on a quantum physical system. Just looking in the box is enough to seal the cat's fate—much more solid evidence of a form of communication than mere telepathy. More recently, though, the imaginary cat has been rendered theoretically safe by the concept of decoherence.

Decoherence is the idea that when a quantum particle interacts with the environment around it, the result is the appearance of the collapse of the waveform. For these purposes the detector and the contents of the box do just fine with no conscious observer required. In effect, the interactions the particle has with the rest of the box have an influence on it. There is a subtle difference between decoherence and a real waveform collapse, but the individual nature of the particle's waveform is effectively tangled up with that of the environment around it, making it behave more like a tennis ball and less like a true quantum particle. It loses some of its quantum weirdness.

There is no absolute answer to the question of what is happening here, at least as yet. The difficulty of untangling the "measurement problem," as it is often called, and saving Schrödinger's cat, remains significant. Attempts to solve this have thrown up complex and messy solutions like the many worlds theory that some physicists believe is the most likely answer. In the many worlds theory, each quantum state exists in a separate parallel universe and there is no collapse, merely a path taken by our reality that effectively moves us from universe to universe. But many others believe this is far too complex a mechanism to be workable, raising many more issues than the problem it attempts to solve.

While a clear understanding of what's going on with the

quantum observation effect escapes us, it does leave open a possible mechanism for telepathy. We need then to get this interaction working across a distance, and that's where the second of the weird quantum phenomena that is often associated with telepathy, the one mentioned by Brian Josephson in the stamp affair, comes in.

Quantum entanglement is one of the strangest aspects of all science—in fact, its remarkable implications were first brought up in a paper that Albert Einstein wrote with two colleagues in an attempt to discredit quantum theory. In this paper from 1935, universally known as EPR after its contributors (Einstein, Podolsky, and Rosen), it was pointed out that there were mechanisms that would make it possible to link together two quantum particles into an "entangled" state. Once they were in this state, making an observation of one particle would instantly provide you with information about the other particle, however far apart they were.

It seemed, Einstein said, that either the distant particle carried the information with it already in some hidden way, or you had to do away with the concept of "local reality," meaning that one quantum particle should be able to influence another instantaneously at any distance. Quantum theory, which dismissed the existence of hidden variables carrying the information, seemed to suggest that information could get from place to place at an infinite speed, apparently running counter to Einstein's special relativity, which limited information transfer to the speed of light. As far as Einstein was concerned, this was the killer blow that would make everyone drop the irritating quantum theory.

We need to understand a little more about what's going on here to see why this quantum entanglement was so remarkable and why Einstein believed he had fatally damaged quantum physics.

Quantum theory was developed in the first half of the twentieth century to explain the strange behavior of quantum particles like photons of light and the particles that make up matter. Although quantum theory emerged from Einstein's own work, he was never comfortable with it. It was because of this theory that he came up with that famous remark:

> Quantum mechanics is certainly imposing. But an inner voice tells me that it is not yet the real thing. The theory says a lot, but does not really bring us any closer to the secret of the "old one." I, at any rate, am convinced that He is not playing at dice.

As we have already seen, the properties of a quantum particle are quite different from the properties of a macro object like a tennis ball. Now here's the remarkable thing that arises from entanglement. Some particles, when entangled, will always be found to have opposite spins when a measurement is made. Measure the spin of one, and if you find it's up, then you will know instantly that the other particle has spin down, even if that particle is on the other side of the universe.

With ordinary physical objects, this is no surprise. Imagine I take a pair of shoes, shuffle them, and randomly select one shoe without ever seeing it. I can then send that shoe to the other side of the universe in a sealed box. There is a 50:50 chance that the shoe I've got back home is left or right. And at the point I look at the shoe back home, and see that it is a right shoe, I instantly know that the other shoe on the far side of the universe is a left. At first sight this is the same situation as with the two particles in entanglement. It's nothing remarkable. But quantum particles are not like ordinary things.

Before I made the measurement, both particles were in exactly the same state: let's say a 50 percent chance of being up, 50 percent chance of being down. There was no hidden information (unlike with the shoes) that determined whether the particle I kept at home would be spin up or spin down. When I then looked at the particle at home and the superposition collapsed into the spin up state, then the other particle immediately became spin down, wherever it was. This fixing of the value did not occur until the moment I looked at my particle—and at that point in time the appropriate information was transmitted instantly to the other side of the universe. This is why Einstein got so wound up.

When Einstein published the EPR paper, entanglement was primarily a theoretical concept. But over the years since, it has been tested and probed many times. Each time, quantum theory has been vindicated. This "spooky connection," as Einstein called it, really does exist. And this is what Brian Josephson was suggesting may be the mechanism behind telepathy. What he was not suggesting, though, was a direct, faster-than-light transmission of information by entanglement.

As soon as people hear about entanglement the first thought is usually that it could be used to build an instantaneous transmitter. The fastest way we know to communicate is using light. This travels at around 300,000 kilometers (186,000 miles) per second. But that still means that a message from the closest star to the sun, Proxima Centauri, would take over four years to arrive—not ideal for conducting a conversation if we ever send an expedition to our nearest neighbor. An entanglement transmitter would overcome this light-lag. But the most remarkable implication is that thanks to relativity we know that a message traveling faster than light would also enable us to communicate with an earlier time.

Einstein's special relativity tells us that the faster something moves, the more time slows down on that traveling object compared to the place it left behind. If we could send an instant message to a ship that had been traveling for some time, so that it had fallen well behind in time, that message would travel into the past.

However, entanglement does not allow for this possibility. Despite being instantaneous across any distance, it does not allow controlled information to be sent. The "signal" communicated this way is always a random value. Over the years there have been many ingenious attempts to come up with an entanglement-based transmitter, but no one has ever devised a mechanism by which it could work. If we take the example of a pair of particles with entangled spin, it's true that once your "home" particle is measured as having spin up we know that the "message" spin down has been sent to the distant particle. But we have no way of controlling what the outcome will be. The choice of spin up or spin down for the local particle is purely random, so all we succeed in transmitting is random noise, not useful information.

I was talking about quantum entanglement at the Edinburgh International Science Festival a few weeks before writing this and an enthusiastic nine-year-old at the front of the audience thought he had the answer to this conundrum. Why not, he suggested, have a series of entangled particles and use their states to send a message. He wasn't the first to think of this. When you take a measurement of an entangled particle's property—its spin, for example—the entanglement collapses.

And it is possible to tell whether or not two particles are still entangled.

So why not have a series of entangled particles as the communication vehicle? At home, at a pre-agreed time, we check the spin of some of these particles, breaking the entanglement. Instantly our distant receiver can then check all the distant twins of the particles. If she treats particles that are still entangled as zeroes and those where the entanglement is broken as ones, she can read off a binary message from home instantly at any distance.

Unfortunately that precocious nine-year-old had missed something that has always undermined attempts to use an extended version of entanglement to send an instantaneous message. While this approach could be used to send a message, it wouldn't work instantaneously. The problem arises when the receiver has to check which particles are still entangled. This is possible, but it can be done only if some information is exchanged between sender and receiver using old-fashioned light-speed communication. By the time the receiver can confirm which particles are still entangled, the time advantage has been lost. There is no instantaneous communication.

This frustrating reality has not stopped many individuals—both scientists and amateurs—from searching for the get-around that will enable entangled particles to truly carry a faster-than-light message. For a while in the early 1980s, for example, physicist Nick Herbert had a design for an instant-communication device based on entanglement in which no one, not even the great Richard Feynman, could spot the flaw. The device depended on the way that photons of light could either be linearly polarized (as with Polaroid sunglasses) or circularly polarized.

Polarization is a way of organizing a property of a photon of light called its phase. The phase changes over time like the rotating second hand of a clock. When photons are linearly polarized, the plane in which the direction of the phase rotates is the same for all the photons. (If you think of the light as a wave, then the linearly polarized waves vibrate side to side in the same direction as one another.) But in circular polarization the direction of the ripple of the wave (the plane of the phase for a photon) changes with time, so as the light moves forward the polarization direction corkscrews around the direction of motion.

Herbert's device relied on sending entangled photons in opposite directions from a centrally placed source. The photon heading toward the receiver would first pass through a laser gain tube, which is a device that would create many copies of the photon. A while later the photon heading in the opposite direction would arrive at a detector belonging to the sender, while the stream of photons heading the other way arrived at the receiver, where the stream would hit a beam splitter, sending half the photons to a linear polarization detector and half to a circular polarization detector.

Now here's the clever bit. The sender takes a look at his single photon, checking for either linear polarization or circular polarization. This instantly forces the remote stream of photons to react differently with the two detectors, because the stream will also become either linearly or circularly polarized. And hey presto, you have instantaneous communication.

Unfortunately for Herbert, ingenious though this idea is, it won't work. This is because you can't magically make copies of photons using a laser gain tube, or any other device. The only way to make perfect copies of a quantum particle, including its exact properties, is by a process called quantum teleportation using quantum entanglement itself, which destroys the original and is anything but instantaneous. A laser gain tube will indeed produce multiple photons with similar properties to the original one, but they won't be identical to the original as Herbert's device required. The process simply can't work. As yet, no one has found the solution.

What all this work on using entanglement for communication, and the clear limitations that apply to it, mean for the consideration of entanglement-based telepathy is that we have to look at something more sophisticated than simply sending a message by examining entangled particles.

What Brian Josephson and others propose is that the mechanism of mind itself involves a large-scale entanglement within the brain. Consciousness is one of the great mysteries of science—we really don't understand how it works. But a number of scientists have suggested that there is a quantum process at its heart, and this may involve significant quantum entanglement within each brain. If this is the case, it may then be possible in such circumstances for two brains to have some overlap of entanglement—in essence, the two individuals' consciousnesses become temporarily linked.

This is a much vaguer concept than Herbert's idea of a communicator based on entanglement, but it is possible that there is something in this as a vehicle for telepathy. The concept then makes use of the idea that consciousness has a practical role in quantum mechanics, as in the impact when the box is opened and the dead-or-alive cat is observed. In effect, the observer is part of the experiment according to quantum theory and can influence the outcome—so it seems reasonable that one mind can influence another by a combination of the quantum observer effect and a linkage formed by quantum entanglement.

Other theories to provide a vehicle for telepathy dabble with multiple dimensions, enabling a connection between minds (or potentially even across time for precognition—see chapter 5) that are physically separated in our familiar dimensions, but that could be directly connected through another dimension. One possibility for such a model for telepathy is to make use of the extra dimensions that are necessary to make string theory work. Variants of string theory do require there to be as many as seven extra spatial dimensions. However, there is a technical problem with this in that string theory assumes—in fact, requires—the dimensions to be curled up so small that they are not noticeable—hardly ideal for reaching across space and time.

A theoretical physicist, Elizabeth Rauscher, has suggested a different approach to provide extra dimensions by extending the normal set of three space dimensions to add three imaginary space dimensions, making six dimensions in total. These dimensions are not imaginary in the sense of "let's pretend there are three more dimensions" but imaginary in the mathematical sense—making use of the square root of minus one.

Generally speaking, a number has two square roots. So, for example, the square roots of 4 are both 2 and −2, because 2×2 is 4 and −2×−2 is also 4. But a problem arises when you try to come up with a number that, when multiplied by itself, produces a negative value. Ordinary arithmetic can't provide the answer, so mathematicians dreamed up i, the square root of −1, making i×i into −1 (as is −i×−i). It might seem that imaginary numbers are just a mathematician's plaything, but they have proved to be incredibly useful in physics and engineering by considering the imaginary number as an extra dimension at right angles to the normal numbers.

Think of a regular number line, which is a bit like a ruler with zero in the middle and all the positive numbers counting off to the right, all the negative to the left. The imaginary numbers can be put on a second number line at right angles to the real numbers. Now any position on the two-dimensional plane can be identified as a complex number—a sum of its value on the real number line and the value on the imaginary number line. So a point may be represented as $2 + 3i$, putting it two units along the real number line and three units up the imaginary number line.

It is this mechanism that makes imaginary (and complex) numbers so useful in physics and engineering, because the mathematics of imaginary numbers has proved a hugely useful tool in making calculations on phenomena that operate in more than one dimension, such as a wave oscillating from side to side as it moves along.

Generally speaking, when the real world makes use of imaginary numbers, there is no problem as long as you don't end up with an imaginary number in the final result. Usually they will be squared or canceled out, leaving only real numbers in the solution. There are some aspects of physics where imaginary numbers seem to lurk with more solidity than most, but at the moment we can't identify anything where there is a real-world imaginary component. In Rauscher's universe each dimension has a true imaginary equivalent—a dimensional equivalent of antimatter—at right angles to it.

Of itself, making these dimensions imaginary does not make them totally impossible. Just as is the case with the description of a wave, the imaginary aspect could just be a convenient mechanism to reflect new dimensions that exist at right angles to each other without interacting. It's just that in this case, each new dimension would have to be at right angles to all the current physical dimensions—hard to get your head around.

The other small problem with Rauscher's model, which would enable two points separated on the normal dimensions to be linked through the imaginary dimensions, is that we have no evidence whatsoever for the existence of these imaginary dimensions. They are a solution to the problem of "How could telepathy work?" not "How can we explain the observed universe?" It would be remarkable indeed if telepathy were the only evidence for their existence—we would expect to see them cropping up in many other aspects of physics. It's all too tempting to suspect that those who reach for other dimensions are merely extending the comic-book conventions of my youth, when strange beings regularly crossed through to our world from "other dimensions."

We are left, then, with entanglement. Even relying on entanglement alone, according to Josephson, there is the potential to provide a vehicle for telepathy. He has suggested that the human mind—and possibly other living things—can latch onto a kind of pattern in what would otherwise be considered a random sequence. Any specific section of a random sequence can incorporate patterns; Josephson suggests that it may be possible for a living organism to latch onto such a pattern within that particular sequence, which gives a possible loophole for getting around the restriction that quantum entanglement can only send random information.

It is just possible that we are looking for too complex a mechanism. Telepathy could be enabled by something more straightforward, dependent on electromagnetic phenomena produced by the electrical activity in the brain. We know that the brain does produce electrical signals and in principle these could be linked electromagnetically to another brain.

This is a little like a remote version of the experiment undertaken by British professor Kevin Warwick of Reading University. Warwick has undertaken a number of studies where he has incorporated small electronic devices into his body. In 2002 Warwick had a chip implanted with a one-hundred-electrode array that was connected to his median nerve fibers, below the elbow joint of his left arm. With this connection, Warwick was able to have some control over both an electric wheelchair and an artificial hand.

Warwick's wife, Irena, was also given a (less complex) implant. With this, she was able to send an artificial sensation to Warwick. A command from her brain activated her implant, which generated a signal. This was then transmitted to Warwick's implant, which finally generated a sensation in Warwick's brain. A communication from Irena's brain was sent to Warwick's brain by electronically extending and connecting their nervous systems. The electromagnetic model of telepathy suggests that it depends on a similar kind of operation, but without the intermediary chips.

The problem with this explanation of telepathy is that it is easy enough to detect the kind of signals that would be required to produce mental electromagnetic communication, and there is no evidence that they exist. As we have seen, sharks *do* have the ability to detect electrical activity in other creatures' nervous systems, and it is possible to imagine that telepathy could be attributed to some extended version of this. But the shark's ability is short-range and limited to simple detection—and it is something we can discover physically when examining a shark's brain.

The human brain would have to be able to broadcast with considerably more power than the very limited electromagnetic signals it does generate to be able to be received by anything other than an electronic sensor in close contact. And while we have clear examples in the shark of biological detectors that can pick up electrical activity, no equivalent has ever been found in human beings.

There have been other suggestions for a mechanism for telepathy. Joseph Banks Rhine (of whom much more in chapter 7), in his detailed study in the 1930s, dismissed the possibility that telepathy could function as some kind of wave-based communication. He had two primary reasons for suggesting this. One was that he believed his experiments showed that telepathy did not become weaker over distances of several hundred miles, making it very different from, say, a radio broadcast.

If this is true, it pretty well rules out an electromagnetic explanation (though it would be no problem for a theory involving quantum entanglement). However, there were a number of issues with Rhine's experiments that meant they did not prove that there was no drop in intensity over distance—and radio operators will tell you that with the right atmospheric conditions a surprisingly weak radio signal can be picked up over such ranges, so this argument does not particularly rule out an explanation based on traditional physics.

Rhine's other argument against any such mechanism was that he was convinced that telepathy and remote viewing (or clairvoyance) were essentially the same thing. In many cases what is portrayed as remote viewing would be better described as telepathy—for example, when an individual is believed to have seen a view through the eyes of an observer. In this case, what is happening is communication between two individuals, rather than literally seeing remotely. Similarly, many supposed examples of telepathy where, for example, a sender looks at a card and the receiver picks up an image of the card could equally be remote viewing, with the receiver seeing the card directly without any intervention from the sender. But if Rhine is correct, being able to detect physical objects without a sender's mind involved makes it harder to accept any physics-based mechanisms.

If a typical picture of how telepathy works is that the sender generates a signal (whether electromagnetic or using other quantum phenomena) that is then detected by the receiver, it is hard to see how a remote-viewing experiment of the sort Hubert Pearce performed for Rhine, where he guessed at card after card in a stack of cards with no one touching them or looking at them (see page 140), could be the result of the same mechanism. As Rhine says, we can hardly expect the ink on the card to generate a signal. Such a feat would need a different kind of mechanism where, for example, some kind of signal from the observer generated a response when it encountered the cards. But again, Rhine's experiments can't rule any possibility out.

When Rhine was writing, basic technologies for remote detection like radar and sonar were still to be developed, let alone the modern equipment we have available. While I am not suggesting that remote viewing or clairvoyance depends on a kind of natural radar mechanism (see chapter 6 for more on remote viewing), if it is possible at all, the lack of physical analogs did mean that Rhine was limited in the ideas that were available to him.

The alternative that Rhine put forward, given these perceived problems with conventional physics, was to say that telepathy (and remote viewing) was liable to have a mechanism that did not involve the kind of physical mechanisms usually studied in a lab. Instead he suggested that the mind was in some way able to dematerialize, disconnecting itself from the body to act independently. The idea was that the mind was able to leave the body and "go out" to the object or sender in order to witness what was required. Rhine seems to have envisaged the mind as a ghostly form, somewhat like the idea of a spirit, that could separate from the body and float around in what he described as a "peculiarly non-mechanistic procedure."

This model of a floating mind clearly builds on a long tradition of "mind-body duality," a concept that dates back to the ancient Greeks but takes its modern form from the seventeenth-century French philosopher René Descartes. The idea, sometimes known as dualism, sees a human being composed of two independent components: a body that is mechanical and material, and a mind (which for religious purposes could also be regarded as the soul) that is supernatural and immaterial. The two are somehow tied together. This is the weakest point of the theory because it requires an interaction between the natural and the supernatural, the physical and the immaterial. Descartes suspected the pineal gland provided the linking point between the two parts of a human being, something we can be sure now isn't the case.

Many people—the majority of people alive today—still hold a dualistic view, because it is the natural, commonsense one. (This doesn't make it right. The natural, commonsense view also says the sun travels around the Earth.) We can't avoid thinking of our "self" as something separate from the body that it controls. We inevitably imagine a sort of mental being, probably located between our eyes, that is pulling the levers to make the physical body work, via the intermediary of a brain.

However, the majority of scientists believe that there is no such duality, and that the mind is simply a function of the chemical and electrical functions of the brain.

It ought to be stressed that this idea of human beings as "meat machines" cannot be proved scientifically, and as yet we have no good answer to the true nature of consciousness. It is also true that to dismiss duality runs contrary to pretty well all the world religions—certainly any that consider us to have a soul or to be capable of an existence after death. However, if there is no duality—and supporters of the mind-as-function-of-brain view would point out that there is equally no evidence for a separate mind other than our subjective feelings—it is hard to envisage how Rhine's idea of the mind temporarily detaching itself—or even projecting itself like a pseudopod from an amoeba—can work.

At this stage, the choice is yours. If you accept that duality exists, you can envisage that there is a possibility of Rhine's mechanism working. If you think that human beings are nothing more than a physical construct of flesh, driven by chemical and electrical impulses, then you have to eliminate the possibility as lacking any real basis.

The final possibility for a mechanism for telepathy is a fifth force of nature. Our best understanding of the universe describes everything as made up of particles that interact through four well-established forces. The most obvious of these forces is gravity (even though it is by far the weakest of the four), while the force responsible for most of the basic interaction of matter—for instance the force that stops your atoms from falling through the atoms of a chair—is electromagnetism.

The other two forces are very short-range, mostly restricted to the nucleus of atoms and nuclear particles like protons and neutrons. The strong force is responsible for keeping together the basic particles called quarks that make up protons and neutrons, and also stops the positively charged protons in an atomic nucleus from flying apart due to electrical repulsion. The weak force is the most obscure, though still essential, as it is responsible for changing the "flavor" of quarks, resulting in the nuclear reactions that power the stars.

As far as we are aware, there are only four fundamental forces, but there is nothing in physics that prevents there from being another. Four isn't a magic number, even though it was popular in protoscience (four elements, four humors). So there could be a fifth force—let's call it the T force, T for telepathy. Forces seem to act at a distance because matter particles exchange special particles called bosons. So, for instance, the electromagnetic force is carried by photons, while gravity is thought to be carried by gravitons, though it can also be seen as a warp in space and time that isn't strictly a force in the same sense as the other three.

If there were a T force, we would expect it to be carried by T bosons, which would cross space, either at the speed of light or somewhat slower, enabling one brain to communicate with the other. The major problem with the fifth force idea, which is popular with some ESP enthusiasts, is that like Elizabeth Rauscher's imaginary dimensions, there is no evidence outside of telepathy for its existing. While it is true that the strong and weak forces were only quite recent discoveries, this is in part because they only work at unbelievably short ranges. We have long been aware of the two forces that act over considerable distances. It seems difficult to believe that there could be a third long-range force that simply doesn't show up in any other circumstances, only in telepathy. Of course, it could be that we have yet to undertake the right experiments, and those T bosons will turn up—but for the moment, pretty well every physicist considers them figments of the imagination.

Leaving aside unknown forces and mechanisms, entanglement may well provide the most likely explanation for telepathy, though it would need a more robust practical theory, going into the detail of what may be happening to be truly supportive of observable facts. But this assumes that telepathy does exist. Is there good evidence to support this to make it worthwhile attempting to produce such a theory, or like those dismissing Brian Josephson's ideas, can we simply say, "Telepathy doesn't exist, so there is no need for a cause"?

One of the first people to claim a telepathic ability was the dramatic, larger-than-life personality Washington Irving Bishop. A successful New York stage performer in the 1880s, Bishop, whose first associations with the paranormal had been working with a spirit medium, claimed to be a "mind reader." This early term for a telepath fit his stage act well. Bishop would leave the auditorium, carefully chaperoned to prevent him from peeking at the audience, and on his return would find objects that had been hidden in his absence, would identify names that had been selected from a directory, and would pick out the "murderer" in a pretend crime that had been acted out on stage.

Rather than claim to be working with spirits like the medium he once worked for, Bishop made a big thing of taking a scientific approach to his "mind-reading" ability. He claimed this was truly a human mental ability. We have no evidence that Bishop cheated by observing the events that took place in his absence, nor did he appear to use a confederate in the audience as other stage performers have done since. However, it is almost certain that Bishop was undertaking a clever stage act.

The method he used seems to have depended on one aspect of his act I have not yet mentioned. (This will frequently crop up as the flaw in reports of apparently wonderful mental acts. There is usually essential information that is held back when describing what happened.) When Bishop returned to the stage and used a (genuine) member of the audience in what was supposed to be mental communication, he would always take hold of her wrist, or connect himself to her by a rigid device like a walking stick.

It may be true that scientists have often been taken in by stage performers, but Bishop's method was spotted by a group of scientists who tested him when he was on a visit to the UK. They suspected that he was picking up small involuntary physical reactions by his subjects, giving away the location of the object or the required person in the directory when Bishop got near to the correct answer.

When the helper was blindfolded it was discovered that he or she could no longer communicate the information to Bishop, something that should have been possible if they were truly linked via telepathic means. The team, which included a leading Royal Society scientist of the day, Francis Galton, also found that replacing a walking stick with a loose chain as the connection prevented Bishop from succeeding, again suggesting he was picking up small involuntary movements by his subjects.

Despite this revelation, Bishop continued to perform to enthusiastic audiences, though his last performance would prove truly bizarre. In 1889, at the age of thirty-three, while performing at the Lambs Club in New York, Bishop was taken ill. By the next day he was pronounced dead, and less than twenty-four hours later he had undergone an autopsy. This was unfortunate, as he suffered from catalepsy, a condition that meant that he could occasionally collapse in a state that resembled death. He carried a card saying that he must not undergo an autopsy until at least forty-eight hours had passed since his apparent demise, in case he was still alive. We will never know if Bishop's last performance was a gruesome dissection of a living person on the autopsy table.

The first big academic name to be involved in telepathy was Joseph Banks Rhine. We will discover much more of Rhine's work in chapter 7, but one interesting aspect of his telepathy studies in the 1930s was his attempts at communication conducted over a distance of 250 miles between Duke University in Durham and Lake Junaluska, both in North Carolina. In one set of tests the results were highly significant, averaging 10 out 25 correct guesses (where 5 would be the expected chance success rate) over a total of 200 trials—not a huge number of tests, but enough to make this many successes noteworthy. Other trials of both telepathy and remote viewing over distances varying from 165 to 300 miles (even between George Zirkle and his fiancée, Sarah Ownbey, who were highly successful when in the same room) produced results that did not vary from the expectations of random chance.

There was one significant difference between the way the successful trial was undertaken and the subsequent events. For the successful trial, the record of the guesses made was sent to the sender, Sarah Ownbey, and she then returned these guesses to Rhine along with the values that she claimed to have transmitted. When sender and receiver independently sent the values to Rhine, the results dropped down to random chance. While this is obviously not definitive proof of cheating, it does cast doubt on the honesty of the recording of the values sent, and whether the values were finalized before or after Miss Ownbey received the guesses. If the values supposedly sent were modified after the event, it would be easy to get whatever score was desired, without in any way doctoring the guesses that had been made remotely.

The problem, as we will see in chapter 7 when we focus on Rhine's work, is that we now know that his test protocols were often not particularly strict, so there is always the lingering doubt that there could have been some form of cheating, whether during the test itself or by subjects altering the recorded results after the test. Also, the run of 200 successful trials ought to have been combined with the other long-distance trials to produce an overall result that was much less significant.

The other issue with Rhine's experiments, something that has been the case with almost all academic work on ESP, is that it didn't attempt to detect what we would normally think of as telepathy. When Professor Xavier in *X-Men* indulges in telepathic communication, a literal worded message passes from brain to brain. The same thing is true of most anecdotal descriptions of telepathy, including my own. But practically every laboratory experiment has used statistical means to look for small changes in perception from those predicted by random chance—and as we will see, this makes it practically impossible to be sure of what is occurring.

There was significant hope of good scientific evidence for telepathy in 1976 when Dr. Charles Tart of the University of California, Davis, published a book containing some striking results. In the book, Tart claimed telepathic test scores that far exceeded the possibilities that chance alone could provide, using an electronic system that he believed removed the potential opportunities for cheating and self-deception that were evident in some of the earlier experimental testing of telepathy. What's more, he believed his system would enable any inherent psi ability to be encouraged by providing appropriate feedback.

Tart's experimental setup consisted of a linked pair of devices (called the "Ten Choice Trainer"), which generated a random choice from ten possible numbers. The randomly chosen value was displayed to a sender in one room by switching on a light next to one of ten playing cards. The sender then attempted to communicate the card he or she was looking at to a receiver in a different room. The combination of a randomly selected target and the separation of the two individuals into different rooms, avoiding any possible visible or auditory clues (whether intentional or accidental), seemed an excellent way to ensure good data was produced.

Unfortunately, there were two big problems with the trial. The first involved the random number generator. If a sequence of numbers (and hence the sequence of the cards being transmitted) is truly random—or close enough for the purposes—then the receiver cannot predict anything about the next number from the previous value. But if there is some way to make a prediction part of the time, even if it only works sufficiently well to give a few unexpected extra hits, the results will be biased in favor of apparent telepathic ability.

Three mathematicians, also based at the University of California, Davis, took a look at the raw data that was used in Tart's trials and identified a problem with the supposedly random numbers that were at the heart of the test. The thing that concerned Aaron Goldman, Sherman Stein, and Howard Weiner was that there hardly ever seemed to be repeated numbers in the sequences. This sounds like a trivial issue, but it would inevitably skew the results.

Let's imagine the random number generator came up with a

value of 2. If the generator is truly random, then the selection of the next number should not be influenced in any way by the previous draw. Randomness does not have a memory—once a value has been selected, then the next value must be chosen totally independently of what came before. The chance that the next number selected is also 2 should be exactly the same as the chance of choosing any other number—in this case 1 in 10. But in reality, this was not happening. There were very few repeated numbers in the sequences used in the Tart device.

This seems to have been a result of a mistake made by the research students operating the trial. To display the next number they had to press a button. If the same number as last time came up, meaning that the same card should be identified for transmission as last time, they tended to assume that they had not pressed the button properly, or that their press did not register—so they pressed the selection button again. Unwittingly they were biasing the output of the random number generator away from randomness.

Of itself, this mistake wasn't disastrous, but it combined with a perception error on the part of the receiver. Even those who are well versed in probability and statistics find runs of the same value in a supposedly random occurrence surprising. This "gambler's fallacy" is why, after several reds in a row, most roulette players assume that the next result is more likely to be black than red, even though the wheel has no memory, and the next win has equal chances of going either way. The receivers, suffering from gambler's fallacy, rarely guessed that there would be a repeated card.

The outcome of this pair of errors meant that the guessing was biased in the same direction as the failures of the random stream of selections. This in its turn would confuse the statistical test used to check for the presence of telepathic ability. This test assumed that there was a 1 in 10 chance of guessing correctly without any knowledge—but in fact the receiver had a 1 in 9 chance of being correct, even if he had received nothing from the sender. He assumed the card was one of the nine choices not sent last time, and that was what he was given. Given that many psi experiments record very small deviations from random chance that become significant only because the tests are repeated many times, this error was more than enough to generate an illusory result.

The link between the poor random sequence and the psi hits was not just a case of thinking that one *could* cause the other—subsequent analysis of the data showed a clear correlation between the degree to which the sequence strayed from truly random behavior and the number of hits that were achieved. Correlation, when two statistics vary in lockstep, does not mean causality. There could, for example, be a third factor that caused both statistics to vary. But in this case it seems highly likely that Tart's results reflected the deviation from randomness produced by the partial predictability of the sequence.

Unfortunately, this statistical error was not the only problem with this particular setup. Tart's device also allowed for the participants to cheat the system, either consciously or unconsciously. Imagine the receiver has just made a choice, selecting one of the ten cards in front of her. She then presses a button on her console corresponding to that card and gets visual and audio feedback of the correctness of the answer. (This is the "training" aspect of the device.)

Meanwhile the sender is watching the receiver on a TV monitor. When the sender sees that the process has been completed, he takes the next random value and pushes a button alongside the appropriate card in front of him. At that point a "ready" light goes on by the receiver—the only communication that travels from sender to receiver other than telepathy, so there should be no way to pass information on. Or so the designers of the system thought.

Yet there is, actually, a second piece of information that travels from sender to receiver, one that can be used to communicate the choice that has been made. This is the amount of time between the receiver making a choice and the ready light going on for the next choice. The time taken for this is controlled by the sender, and is communicated to the receiver. If there is collusion between the two, it will be easy enough to set up a coding system where the amount of time indicates which card has been chosen. But even if there isn't collusion, if the sender consistently puts in different delays corresponding to some of the cards, then it is entirely possible that over time the receiver will unconsciously pick up on this and find that her scores have improved.

Such an unconscious difference is quite likely to occur. If face cards are used, it might be harder to concentrate on these than on a simple number. Similarly, if the cards' values run from 1 to 10, some numbers might take longer to visualize, a possibility that seems entirely possible when you bear in mind the approach taken by Tart's best sender, one of his students, who described fixing an image of the card as if it were floating in front of his head before pressing the button that lit up the ready light. It's entirely possible that this fixing process took longer for cards with large numbers of shapes on it that were more difficult to visualize.

When Tart repeated the study with the potential defects removed, there was no sign of telepathy—the guesses using the same device but with appropriate improvements in protocols were no better than you would expect from random chance. Bizarrely, rather than accept that this was good evidence for the flaws in the design of the original experiments, Tart's account of the second study suggested that the results did not make it necessary to discount the first study, as the original results were so good—hardly significant if the experiment that produced those results was fatally flawed.

These experiments directly transmitting values, and the straightforward experiments transmitting simple images on cards used by Rhine (see page 162), have largely fallen out of favor, and beginning in the 1980s there was a rise in popularity of an experimental approach to detecting telepathic abilities given the impressive sounding name of "ganzfeld." Before we take a look at this, which is considered by many to be the most scientific of the telepathy tests, it's worth taking a brief excursion into dreams.

Dreams are a natural subject to consider when dealing with psi. Their unworldly oddness ties in well with the "spooky" aspects of psychic phenomena, and dreaming seems to be when our minds run most freely—perhaps making them most open to telepathic communication. A series of tests of dream telepathy was undertaken over a period of ten years beginning in the early 1960s at a dream laboratory set up by Montague Ullman at the Maimonides Medical Center in Brooklyn.

A typical test involved a sender, a receiver, and an experimenter. The experimenter would monitor the receiver, who had the never easy job of falling asleep under observation in the lab. When the receiver entered the rapid eye movement (REM) sleep that accompanies dreaming, the experimenter signaled to the sender, who opened a sealed envelope and took out an image that would be used as a target for the telepathy transmission. At the end of the REM period, the receiver was woken up and reported his dream, then went back to sleep. Each time the receiver went into REM sleep, the process would be repeated with the same image.

In the morning the receiver was asked to choose between a range of pictures, typically between eight and twelve very different images, to see which best matched his dream recollections. Then a panel of judges would also try to match the recording of the dream logs to an image that best corresponded to the description. Some, but not all, of the tests came up with a significant statistical bias toward the correct answer, though this was never anything as major as, say, getting it right half the time. As is often the case with such tests, we are talking about a low numbers of hits, and in this case, because of the intensive nature of the trials, taking three people for a whole night to complete a single test, the experiment featured a relatively small set of trials.

The positive thing about these experiments is that if the protocols were adhered to well and there was no conscious or unconscious bias from the experimenters, the choice should have been blind, on the part of both the receiver and the judges. The difficulty, apart from being sure from anecdotal evidence that the controls were strict at a time when many other experiments certainly lacked decent protocols, is the vagueness of the targets that were used. These were pictures of artworks.

To illustrate why this is a problem, one of the Maimonides tests involved the transmission of Max Beckmann's painting *Descent from the Cross,* showing Christ being taken from the cross after his death. The opportunities for potential links here in a culture with many Christian references are many, as is made clear by the fact that one of the indicators of the receiver's success in this particular trial was having a dream featuring Winston Churchill. The wonderfully tenuous link here was church (hence Christianity) and hill (as in Golgotha, the hill on which Christ was said to be crucified). Presumably if there had been a dream about someone called Peter or Mary, or about carpentry or woodwork or religion or death or execution or Romans or Israel . . . in fact, pretty well anything, it would have been counted as a success.

When not writing popular science books I give training in business creativity. The aim of the creativity techniques I teach is to free up the mind and allow it to generate as many new ideas as possible. In twenty years of experience in creativity I have discovered one technique that is more powerful than any other at producing a vast range of possibilities. In the technique, I use a randomly selected picture as a stimulus to help participants come up with new ideas. As a simple demonstration of the effectiveness of this, I have used the same picture to inspire trainees to come up with solutions to the same problem in every introductory session I have given—which adds up to several hundred runs of the test. Each time without fail someone has come up with a totally new idea, despite using the same picture as stimulus for the same problem.

What this demonstrates is just how powerful a picture is at generating a wide range of associations and links, the essential building blocks of ideas. If you combine the complexity of a picture with the rambling, often lengthy nature of a dream narrative with lots of detail, it is almost certain there will be a fair number of correlations, because that is the nature of an image of this kind. It is excellent at diverse associations. This vagueness was unforgivable in the Maimonides experimental design, which should have restricted the subject matter to something simple that would have provided a straight yes-or-no answer and could have had a huge range from which to select the random choice, such as a single word or number. Even though the experiment required a match to one from a choice of pictures, the process was simply not clear enough.

Returning to the ganzfeld experiments, which remain the most discussed recent telepathy experiments with any scientific basis, we begin with something that sounds more like a sci-fi mind probe scene than a parapsychology experiment. The receiver is isolated as much as possible from any distracting structured external sensory input. Table tennis balls cut in half are taped over the subject's eyes, with a red light playing on the plastic surfaces. Pink noise (white noise with the higher frequencies filtered out, which makes it more pleasant sounding) is played through headphones, and the receiver lies back in a reclining chair.

This disturbing-looking technique was originally developed in the 1960s as a way of providing limited sensory deprivation, but it was picked up in the 1970s by a number of parapsychologists. The idea was that the technique would minimize external conventional sensory stimuli that carried any information, enabling the subject to focus on any inner stimuli—in this case a telepathic communication—rather in the way that the deprivation of sight tends to intensify a subject's ability to hear. This odd methodology was used rather than simply totally cutting out sensory input because it was felt this would help the subject concentrate, whereas a complete deprivation of sensory input would tend to encourage her to drift off to sleep.

When the receiver has settled into her isolated state, the sender is given a randomly selected image and concentrates on it. The receiver spends some considerable time speaking aloud her sensations, typically around half an hour; at the end of this period, the receiver is given four images to look at: one showing the target and three others. Joseph Rhine would point out that a flaw with this test is that it can't distinguish between telepathy and clairvoyance, but in the end the outcome would be significant in either case, and the distinction between the two would be of any interest only if there were positive results.

The expectation with four images to choose from is that one in four guesses would be a success by chance alone—as usual with such experiments, it is only with large enough samples to enable good statistical validation that the outcome can become meaningful. One obvious problem here, compared with a quick guess of the values of a pack of cards is that ganzfeld is a very slow process. A traditional card-guessing technique could get through a hundred or more tests in the time that a single ganzfeld process takes. In a summary of ganzfeld experiments made in 1994, a worryingly large number consisted of fewer than twenty trials. Compare this with the thousands of card trials undertaken by Rhine in the 1930s.

Looking back over the ganzfeld trials, pretty well everything that could have been done badly from the point of view of good experimental protocols seems to have occurred at some point. Often several statistical methods were tried to analyze the data, with the method providing the "best" result being selected. Unsuccessful studies were likely not to be reported (a phenomenon so common it is given a name: the file drawer problem), and there was a tendency to be selective about results, so a series of exploratory trials would be counted if successful but ignored as exploratory if not. And once again the information to be transmitted was often a complex image or even a video, introducing a painful and subjective vagueness into the matching process.

Yet another problem with ganzfeld was that without specifically being designed to do so, it provided the magician's favorite technique, misdirection, in spades. We get the impression from the impressive sci-fi look of a ganzfeld receiver, bathed in red light, wearing earphones and Ping Pong balls, that she was impervious to outside communication. And she was pretty much while in the chair (though the headphones could raise a suspicion of carrying radio communications if cheating was suspected). But the way a magician makes a trick fool us is by emphasizing how secure controls are at on one point in time and space, while actually manipulating things at another.

I am not suggesting that the ganzfeld experiments were consciously rigged, but there was a much better opportunity for communication by conventional means at the point in a test when the receiver had been stripped of the weird-looking apparatus and was given the four images to choose from. If anyone who happened to know the right answer was present when the images were shown to the receiver, it would be easy enough to accidentally communicate this, just as Washington Irving Bishop's audience members unconsciously communicated what had happened when he wasn't on stage.

In some, but not all, versions of the ganzfeld experiment there was also a clear experimental error at this point because the same physical picture that had been used by the sender was included in the pack of four presented to the receiver, rather than using a duplicate. There are several reasons why a picture that has just been handled for thirty minutes could differ in appearance and feel from the other three that have not been handled, from crumpling of the paper to smudges from fingerprints. Whether the receiver was conscious of this or not, she could single out that image as the one that felt different from the rest. Because it *was* different.

The part of the experiment where the receiver made a choice from four (why only four?) possible images was, without doubt, the weak point of the experiment, where there could easily be a failure to keep conventional information away from the receiver. Once she made the correct selection, it really didn't matter what she had spouted in her thirty minutes of contemplation in the ganzfeld chair—human beings are very good at rationalization after the fact and would find a way to link the rambling recorded words with the image that was selected. The utmost care was needed to ensure that there was no unintentional communication at this point, but it seems that this care was not exerted when most of the experiments were undertaken.

In these statistically based tests the shift from the outcome of random chance is often tiny, so it takes only a marginal influence to produce the desired effect. Imagine, for example, that the way the target image was added to the pack of four was not randomized but either fell in some sequence or was always (say) the top or bottom image in the stack. This is all it would take to trigger statistical significance over a good number of trials.

Take another possibility where an error could emerge from an innocent statistical effect. In the better-run trials, images were selected using random numbers to avoid there being any potential prejudice in the way the targets were selected. But let's imagine that the sequence of random numbers used happened to select some images more often than others. This is entirely possible with a relatively small sequence of random numbers. Bear in mind that you would expect to see repeated values sometimes in truly random numbers. The nature of the ganzfeld experiments, taking such a long time over each test, tended to encourage relatively short sequences and hence increased the possibility that any particular random sequence would actually contain a pattern.

If one of these patterns cropped up, some images would be used as targets more than others. Unfortunately, people have preferences. When an individual is looking at an image and trying to identify the one that "feels" right, he is more likely to choose certain images rather than others. Those preferences may not be consistent from person to person, but any subject is likely to have favorites. All we need is for the subject to have a very slight preference for an image that happens to have come in the random sequence more often than others and immediately we've got more than enough of a bias, purely from guesswork with no communication, to appear to have a significant result.

This kind of error is not unavoidable. One obvious possibility is to have a vast number of images and never to repeat them, though even then you would have to be careful not to have a bias toward certain kinds of image. Ideally, before the ganzfeld test you would put the subjects through another test that established their personal weightings toward various images, and you would use only a set that came up with neutral weighting. It would also be much better, retrograde though it may seem, to revert to simpler images like those on the old Zener cards (see page 162), which featured basic shapes, though with many more variants. These may well be more effective than the complex images typically used in ganzfeld, where it would be difficult to clearly identify and iron out personal preferences.

Given the concerns about the earlier studies using the methodology, later ganzfeld trials were undertaken trying rigorously to avoid the potential problems. A summary of these trials through the 1990s showed that with the extra controls there was hardly any deviation from the expectation of chance. Others have attempted a more rigid link of output from the receiver with the input in what have sometimes been called autoganzfeld trials. In these, the sender watches a video and the receiver gives a running commentary of what he feels is being watched. The commentary and the events in the video are then matched together, looking for hits.

There are two issues with the autoganzfeld approach. One is that it is an expensive and laborious process, far more so than the original, already clumsy ganzfeld method. It is necessary to go through and check around half an hour's video and commentary second by second looking for correlations, and then to put the whole through some kind of statistical analysis (exactly what kind of analysis is appropriate causes considerable argument). This means that it is difficult to get successful experiments reproduced and there tend to be very few trials.

The second problem is the "inevitable coincidence" effect we have already seen in an artwork but multiplied by all the different scenes that occur in even a short video clip. Practically everybody has experienced what appear to be unlikely coincidences in their lives. Some of these are "seeded" coincidence. Consider this classic psi example from the early days of parapsychology. You are walking along a road and think of a tune, as we often do. Someone else comes down the road in the opposite direction. Not only is the other person humming the same tune, but he is at almost the same point in the tune: a great example, surely, of telepathy occurring in the wild.

Only this particular tune happens to be popular. You retrace your steps and hear the tune on a radio, the sound coming from an open window. Along the route, now that you are conscious of it, you heard the same radio station several times, floating from windows. (This example was first thought up back when there were very few radio stations in the UK.) Both the "sender" and the "receiver" have picked up the tune unconsciously, hearing it from a window.

However, there is no need to have such a seeded coincidence—genuine, real coincidences happen all the time. Take a simple measure: meeting someone you know in an unexpected place. I can think of at least four times this has happened to me. Once I was crossing a road in a village around two hundred miles from Cambridge and another member of the same small Cambridge choir I belonged to walked across the same street in the opposite direction. On another occasion I got off a train at Oxford station, and standing directly opposite me on the platform was someone I knew from college, many miles away.

I have also bumped into work colleagues, once on a tube station in London when neither of us worked in London, and once at an airport in Germany. These coincidences feel remarkable, though there are usually factors that make them more likely to occur than pure random chance. However, what we don't notice are the many, many millions of moments when we don't bump into someone we know, or whatever the coincidence happens to be. Let's face it, you aren't going to impress an audience in a bar by telling them the story of the day you went out and didn't see anyone you knew in a surprising location. But that's what happens practically every day.

The autoganzfeld experiments suffer from the same problem. When you put together two detailed streams of occurrences—what happens in the video and the verbal description given by the receiver—there are bound to be some remarkable coincidences.

In one experiment, for example, there was a scene of someone falling over in a stark landscape, while at the same time the receiver remarked that she was picking up that "someone falls hitting their face on stony ground." Perhaps, this isn't quite as dramatic as meeting a friend in the middle of nowhere, but it is still striking. But the very nature of the experiment opens up the potential for many coincidences like this. What would be even more remarkable is if there had never been any coincidences at all.

Because a video is being used rather than, say, a number, a word, or a simple symbol, there is a very rich collection of information being focused on by the sender. Similarly, the receiver reels off a huge amount of impressions. If, between the two, it isn't possible to get hit after hit it would be quite surprising.

There seems to be little advantage from the ganzfeld process, while the limitations of taking so long over each trial mean that it is difficult to gain sufficient data. Particularly in the autoganzfeld format it is also the case that what is being processed is simply too complex a set of information—a similar problem to that experienced in many remote-viewing experiments (see page 128). To demonstrate effective results we should be looking for clear, simple pieces of information that are either right or wrong.

By using experiments involving such rich and complex data, I believe, the experimenters made a huge mistake, as it became impossible to separate noise and signal, and unwanted coincidence was inevitably introduced. It is also unfortunate that they didn't run control experiments where no one was looking at the image (distinguishing telepathy from remote viewing) or where no image was selected until after the results had been assembled (without telling the receiver); control experiments are hugely important in a scientific experiment, and their absence is telling.

For a final example of experiments attempting to discover telepathic ability (though we will come back to them in later chapters), we need to visit the Physics Department of Boston University in 1977. While it's not entirely clear whether he was attempting to test for pure telepathy or for a combination of this with telekinesis, one of the strangest attempts to detect mental capabilities came from a researcher who would be heavily involved in experimental work on quantum entanglement in the United States, Abner Shimony.

As we have seen with the Schrödinger's cat thought experiment, theorist Eugine Wigner had suggested that the quantum state of a particle was established only when it was observed by a conscious observer. If this was true, Shimony thought, it should be possible to use this effect to provide a very sensitive detector for telepathic ability that totally removed the vagueness of the kind of person-to-person communication that the usual experiments looked for.

In his experiment the sender was in a room with a source of radioactivity and a detector, while the receiver was in another room. The theory was that by choosing whether or not to look at the display of the radioactivity detector, the sender would be able to influence the state of the quantum system that included both him and the equipment, producing an effect that the receiver should be able detect when he looked at his own display attached to the same detector with a delay line.

The result was unsuccessful, coming up with a pure chance outcome, though it is not entirely surprising given the combination of a very speculative interpretation of quantum theory with an equally speculative assumption that the change in the state of the system would result in a message being received. Even so, it was an interesting experiment with which to leave telepathy for the moment. It's time to get away from the relatively subtle idea of influencing another mind and go for the big one. Full-blown mind over matter. Telekinesis.

4. IT MOVES!

In the center of a table that has been carefully checked for rigged devices sits a magnetic compass. The psychic takes a seat at the table. "With the power of my mind alone," he says, "I am going to make that compass needle move." He waves his hands over the compass, and as he does so, the needle begins to twitch, following the movement of his hands as if they were magnetic. You checked before that he wasn't holding anything, and he has rolled up his sleeves. He wears no rings or watch—but you are still suspicious. Perhaps he has metal dust under his fingernails.

You run a second compass around his hands, but there is no deflection of the needle. If he had magnetic material on his hands he has already got rid of it. "I can see you are careful with your controls," says the psychic. "That's fine—I like to make it very clear what is happening. Look, it's much harder for me to do it this way, but I will give it my best shot." He places one hand on either side of the compass. "I will attempt to deflect the needle without my hands moving at all," he comments.

Not wanting to be caught out, you stare at those hands for the smallest movement. The psychic's face distorts as he makes the attempt. Nothing happens. He tries a second and a third time. Still there is no result. You can see the muscles straining in his hands as he forces them to keep still. It looks like it is going to be a failure. The psychic starts to say something as if he is giving up. He is not even looking at the compass needle when finally it moves again. Even he seems shocked when one of the witnesses notices it. "Look!" he shouts excitedly. "Look what is happening!" Surely you have just witnessed a definitive demonstration of the power of the mind?

Many, many people at some time in their life have tried telekinesis—attempting to move things remotely using only their mind. There is something particularly appealing about the idea of thinking at something and making it move. Whole generations of moviegoers have grown up with *Star Wars* and the idea that it is possible to use some kind of mental force to think something into action, whether it's Yoda lifting an X-wing out of a swamp or Jedis mentally summoning their light sabers. We might not have the Force, but it feels right that we only have to exert enough mental effort and things will begin to move.

There are many ways to fake telekinesis—stage magicians do it all the time—which makes it particularly dangerous when demonstrating it, as Uri Geller and many others have, using a device that responds to invisible forces, like a compass. The thing about magnetism is that the source of it is not obvious. The compass itself demonstrates this all the time. Where does the magnetism that makes the compass needle point to the north come from? We know it's the magnetic field of the Earth—but it's certainly not obvious that this is the case. So it's hard to think of a worse way to demonstrate telekinesis.

In the imaginary demonstration above, based on real examples, the psychic was cheating—but in such a way that watchers would be fooled, as so many observers have been in the past. The whole effort was designed to focus the audience's attention on the psychic's hands. We made sure that those hands were controlled. There could have been even more controls—for instance, his hands could have been fastened behind his back with handcuffs. But there were at least three other places that he could have had a magnet or a piece of metal concealed: in his mouth, in his shoe, or in the knee of his trousers. In each case, while you were focusing on the hands, he could have used another part of his body to give the compass the invisible kick it needed to appear that it was being moved by telekinesis. Misdirection was a hugely important part of this magic trick, and is always a common factor in fraudulent psi demonstrations.

The table, a natural enough part of the setup to keep the compass stable, helped hugely. For example, our performer could have brought his foot up under the table to position it close enough for the magnet to have an effect if it was in his toe cap. Or the magnet might have been in his mouth. We all know that psychics tend to writhe around while they are trying to make something happen. But because all our focus was on his hands, and he kept them still, we didn't connect the movement of the head and the jerk of the compass needle. Particular care will be needed when looking at something so easily duplicated by natural forces as telekinesis.

Unlike telepathy, in telekinesis there is something more substantial than information being transmitted. Telekinesis requires a force to be applied to an object. This implies the transmission of a considerable amount of energy from place to place, combined with the need for basic physics to be considered. Where, for example, does the equal and opposite reaction of Newton's third law go? According to Newton, when we push something, there is an equal and opposite force pushing back on us. If a mental force is applied to something, what does the equal and opposite effect act on?

One essential to make thing happen is energy. The laws of thermodynamics, which govern the transfer of energy between systems, are among the most well supported in all of physics; yet if telekinesis exists, the effect seems to bend, if not break, the laws of thermodynamics. Where is the energy coming from to move the object? *Star Wars* characters can use an imaginary all-pervading Force, but for telekinesis to work for real we need a real source for that energy. It could in principle be our brains, which take a fair amount of energy to run—about one-fifth of the roughly hundred-watt output of the body—yet very little of that energy seems to be leaking out other than as heat. To make macro-scale, visible telekinesis work, the energy would have to either come from the environment, cooling down the molecules around the object being moved, or be transmitted in some way from the brain.

It's true that electromagnetism can transmit a degree of energy across space, but the level of that energy drops off very quickly, which is why for an induction transmission system like the one used to charge an electric toothbrush, the brush is practically in contact with the charger. It's true that you can transmit power farther across space. A dramatic demonstration of this can be made at night using a long fluorescent tube. Take it under a high-voltage overhead power cable and plant one end of the tube in the ground. The bulb lights up with no wires attached, powered by the voltage difference between the end that is grounded and the other, a voltage difference induced by the power line.

Such transmission of power was the dream of the electrical engineer Nikola Tesla, who believed it should be possible to broadcast power across hundreds of miles through the air. But Tesla's dreams were never made a reality, and to induce a current at a distance requires huge voltages, quite different from the tiny electrical charges in the brain. It is hard to see how there could be an electromagnetic explanation for the transfer of energy required to put something in movement. The character Magneto in the X-Men movies and comics achieves a kind of telekinesis by manipulating huge electromagnetic fields—but unfortunately, outside the comic book world this kind of electromagnetic power is not just waiting to be tapped.

Some psi researchers believe that telekinesis is the result of a fifth force of nature (see page 55), but it seems unlikely that this would not have been discovered in other ways. Others get around the apparent lack of a mechanism by taking telekinesis to such an extreme that it takes a stretch of the imagination to make what they are studying have anything to do with mind over matter. As we will see in the section on the PEAR lab at Princeton, some attempts were made to perform telekinesis at the quantum level where the same kinds of quantum mechanism suggested for telepathy could change the quantum state of an electronic device, producing, for example, a shift in a random number generator.

This is hardly the kind of thing we typically think of as telekinesis, though. As yet I have not seen a single viable mechanism for macro-scale telekinesis to work. This doesn't mean it doesn't exist. I haven't seen any very convincing explanations of the mechanisms of the dark energy that is causing the universe to expand, but that doesn't mean that the expansion isn't taking place, nor does it mean that we can't discuss it. So what has been observed in attempts at experimental telekinesis?

One of the earliest scientific attempts at telekinesis used what seems a very sensible approach of trying to move a very delicate chemical balance, designed to measure thousandths of a gram. This was undertaken by the physicist and psychic enthusiast Sir William Crookes, now probably best known for the Crookes radiometer toy, where a series of paddles rotate inside a sealed bulb because the paddles are reflective on one side and absorbent on the other, so tend to warm up differentially and receive more pressure from the diffuse but present air molecules on one side of the paddle than they do on the other.

By a strange coincidence, because it moves without being touched, the radiometer is quite often used in amateur tests of telekinesis, though it is very difficult to undertake such a test without changing the light falling on the device and hence influencing it through natural means that have nothing to do with thoughts. Strictly, a test version of a radiometer should have both sides of the paddles painted the same color and should be tested under powerful lights to make sure that it has no natural inclination to turn on its own.

Crookes strained manfully in his attempt to displace the chemical balance and failed. When Joseph Rhine's lab (see page 140) attempted to look at telekinesis, for some reason the researchers used a much more indirect mechanism, which, like their attempts at telepathy and clairvoyance, would rely on a statistical outcome to determine whether or not the effect was present. While this is an understandable convenience with telepathy, it is harder to see quite why the experimental procedure of moving something extremely light wasn't adopted. That way a test could be much more definitive than the approach actually used.

Instead, Rhine's test involved influencing the outcome of dice throws. Thinking of a comprehensible mechanism for this to work is much harder still than only moving something with the mind. The mental manipulator has to somehow be aware of the way a die is rolling so that he or she can influence it to be more likely to fall one way or another. This seems to require a huge degree of mental perception of the position of the die and an ability to calculate its dynamics on the fly on top of the ability to move it, a much more complex psi capability—if this were possible, it would truly be a superhuman capability.

In fact, Rhine's tests did demonstrate on various occasions that a person could influence the outcome of a dice throw. Casinos around the world should have been trembling with fear. Their lack of concern should be a good indicator that human beings can't influence dice throws. If all the years of millions of people attempting to mentally influence the rolls of dice in craps games didn't leave the casinos at a loss, it's hard to imagine that it is possible to bias dice this way. However, the casinos needn't have worried. Sadly, as often seemed to be the case with the Rhine tests, the design of the experiment had a fatal flaw. Most of the tests did not have a control run.

Consider this actual experiment undertaken at the Rhine lab at Duke University by a graduate student called Frick. He threw a die (with some dedication, it must be admitted) 52,128 times while trying to mentally influence it toward a six. According to pure chance there should have been 8,688 sixes thrown. In practice the result was 9,720 sixes—an extra 582, which is far too many to be accounted for by chance alone. It seems that Frick really could influence the dice and should have been on his way to Las Vegas to clean up and paint the town red. But unlike many of the other comparable Rhine trials, Frick undertook a control run, a process that should have been absolutely essential in all such tests.

In the control run, Frick made exactly the same number of throws, but this time he either attempted to get a one or merely tried to avoid getting a six. This was an important addition to the experiment because without it, the encouraging results he produced could have been attributed to a bias in the die. When you think about the way the dimples are distributed on the face of a die, it isn't surprising that they often tend to have a small bias toward the six. The side with six indentations is always opposite the side with one. So the lightest side (with more wood or plastic removed in the extra dimples) is opposite the heaviest. If the die has a slight tendency to end up heaviest side down, it will be with one at the bottom—showing six on the top face. Purely coincidentally, people more often than not try to influence a die toward a six, presumably because it's the "best" score, if given the choice in these experiments.

What was the outcome of Frick's control trial? This time he got 9,714 sixes—again exceeding the expectations of random chance, in this case by 576, a very similar value to the first attempt. Bizarrely, when Frick's results were revealed, rather than accept that the die was biased (something that these days we would expect to be tested by using an automated shaking device with no one present), Rhine and his colleagues assumed that Frick had made an error.

Their argument, worthy of Monty Python, goes something like this: As soon as you are told not to think of something, it becomes impossible not to. Try it. Spend thirty seconds not thinking of a polar bear. How did you do? The sheer need to know what it is that you aren't thinking of means that you inevitably do think of it. Rhine and colleagues argued that Frick was unconsciously still influencing for a six in the control series. The very fact that he was trying not to think of six made sure that he did. This argument is all very well, but it is hard to see why this was a problem when he was trying to force a one, rather than the admittedly difficult feat of *not* thinking of six.

What is beyond doubt is that telekinesis is a favorite of stage magicians, and special care needs to be taken to separate the individual performing the feat from the object that is to be moved so that there can be no physical connection between the two. At the very least it would seem that a glass screen would be essential, because it has been well documented that dramatic effects can be produced by using a carefully focused jet of air from the mouth. What can seem a remarkable demonstration of mind over matter can often end up as nothing more than an impressive display of concealed blowing—clever and skillful, certainly, but not in any sense paranormal.

This seems to have been the specialty of stage performer and ex-convict James Hydrick, who had two signature telekinesis tricks that he performed many times in the early 1980s: making a pencil rotate when it was hanging over the edge of a table, and getting the pages of a book like a telephone directory to turn without touching it. I have seen this on video and it does look very impressive.

Since Hydrick started his act, it has repeatedly been demonstrated that both of these abilities can be reproduced by stage magicians if considerable steps are not taken to prevent the performer from blowing—in fact, magical performer and professional skeptic James Randi has even demonstrated page turning by blowing a stream of air through a narrow gap at the base of a bell jar that apparently totally protected a book from air currents. It was interesting that when Randi scattered a few extremely light Styrofoam chips on the book pages, which would have been wafted off by any attempt to shift the pages by blowing, Hydrick was no longer able to perform his act.

A magician with a genuine belief in some aspects of parapsychology, Danny Korem, filmed Hydrick doing his act. Korem had already practiced the techniques required to issue undetectable controlled puffs of breath and immediately duplicated Hydrick's performance using nothing more than a stream of air. Other, perhaps more self-confident, performers like Uri Geller might simply have congratulated Korem on his skills but pointed out that theirs were totally different. Just because a magician can duplicate the effect of a magnet, for instance, doesn't mean that electromagnetism isn't real and magnets don't exist. Hydrick, though, confessed all, admitting that he had faked these abilities as an attention-seeking move.

There is one point that needs clearing up here, given that Hydrick, like Geller before him, was a stage performer who claimed to have genuine psi abilities. Isn't it pointless testing someone who makes money out of performing these kinds of tricks for an audience? Doesn't that inevitably make them fakes? This was certainly the opinion of a scientist I saw a number of years ago on a TV documentary on paranormal abilities. "Surely," he said, "anyone with such abilities would put himself forward as a research subject, rather than go on stage." For me, though, this statement only demonstrates the unworldliness of the scientist (I can't remember who the scientist was, though it may have been Richard Dawkins). Think of the choice that a true psi performer would have to make: become famous and make lots of money, or go and have your brain experimented with in a lab. If you had such an ability, which option would you prefer?

In the end, telekinesis stretches the imagination to the breaking point because it requires a link between mental effort and a remote physical piece of work that requires a significant amount of energy to be expended. We need to be very clear that there is some mechanism enabling that energy to be transmitted, and for the movement to take place with all the basic physical implications that go with it. But at least telekinesis does not break the boundaries of time, potentially disrupting the linkage between cause and effect. The most dramatic of the psi abilities, precognition, suggests that the human mind can see into the future and detect events before they have even occurred.

5.

THINGS TO COME

On May 3, 1990, I was due to catch a flight from London to Las Vegas to attend a trade show, as part of my work on new technology for British Airways. As an airline employee I had always flown regularly and had enjoyed the experience. Just that Valentine's Day I had flown to the most extreme ends of the British Isles—the Shetlands to the north of Scotland and Jersey in the southerly Channel Islands off the coast of France—just to post a series of mystery valentine cards on behalf of friends and relations. I didn't give a second thought to stepping on a plane.

The night before I was due to fly to Las Vegas I woke up in a cold sweat. I was convinced that my flight the next day would crash. I felt that I had foreseen the future. As someone with a scientific background, I had no belief whatsoever in precognition, the ability to see what has yet to be. I was uncomfortable with the concept of prophecy. In my head I knew that what I was experiencing had to be without substance. Yet I was convinced in my gut that if I got on that flight I would not make it across the Atlantic.

A caring boss allowed me to take a different flight—and of course nothing happened to the original plane I should have been on. There was no crash. But that sensation of knowing what was going to happen was remarkably strong. I *knew* that a terrible story was about to unfold. Because nothing did happen, I didn't mention it to anyone. Until now it is not something I have talked about (except to my boss). But what if something had happened? Then, certainly, the information would have got out, would have been made public. After the event I could have told everyone that I had made a chilling prediction, and I even had a witness to prove it.

And here is the first problem with precognition: there is an inevitable self-selection of the anecdotal evidence for its effectiveness. Lots of people think of something happening—good or bad—in the future, but usually it doesn't come true, and they soon forget their premonitions. But in the unlikely event that something really does happen, the apparent link will be brought out. Just think of the millions of people with a lottery ticket who have that moment of delicious anticipation when they hope they have won something, only to be disappointed. But someone has to win, and for him, the "precognition" of the win is a story worth telling. All the millions of others whose feelings let them down will not come forward to tell about the day they thought they were going to win the lottery . . . and didn't. Yet we are amazed when one person in a million happens to think of something in advance and it really does take place.

Prediction, as Yogi Berra, Niels Bohr, and others are alleged to have said, is difficult, particularly about the future. The psi ability of precognition is an awareness of what is about to happen, being able to see what is coming. There have certainly been many who claimed to be able to forecast the future, whether Nostradamus and his intensely vague prophecies or the predictions of weather forecasters, but most of these are engaging in guesswork about what is going to happen, basing the guesses on either gut feeling or scientific models. Precognition is different—it is the ability to see accurately into the future.

It might seem that this immediately rules out the possibility of precognition. We are used to the future being considered an "undiscovered country," in part because this phrase from Shakespeare was used as a subtitle for the sixth *Star Trek* movie, though in fact Shakespeare did not say that the future is an undiscovered country, but that death is. The past has happened and is set in stone, it would seem, but the future has yet to happen, so it is made up of a myriad of possibilities in a shifting chaotic pattern.

However, scientists learn never to say "never." There is nothing in the laws of physics that prevents time travel, and in principle there are mechanisms that would allow information to travel into the past from the future, giving someone forewarning. But such mechanisms require us to invoke relativity. The easiest way for information to travel into the future is to send it off at very high speed in a spaceship. But to get anything into the past requires general relativity and usually large-scale manipulations of gravity—a nontrivial possibility at best.

It is hard to see how you can go from any concept of what a practical time machine could be like—usually involving the manipulation of neutron stars or hypothetical wormholes in space—to the way most precognition is performed by merely thinking about the future and getting a feel for what is going to happen. One is a major manipulation of the physical universe. The other is just getting in the right mental state.

Perhaps the closest scientific theory to one that might provide an explanation for precognition is that of advanced waves. This concept emerges from Maxwell's equations, the mathematical formulas that elegantly describe the nature of light as an interaction between electricity and magnetism. Like many equations, Maxwell's have more than one solution. Just take a simple example we have already met:

$$x^2=4$$

What is x? Your immediate response is probably 2, because $2 \times 2=4$. And you would be right. But as we have seen, an equally valid solution would be −2, as -2×-2 is also 4. Similarly, the equations describing the behavior of light have two equally valid solutions. One produces what are usually called "retarded waves"—these are the waves of light as we know it, the ones that leave a light source and depart for their destination. But the mathematics provides a second solution, known as "advanced waves," which leave the destination and head back to the source, traveling backward in time.

With the possible exception of precognition, we have no direct experimental evidence for advanced waves existing, and for a long time they were just considered to be a peculiarity of the math and simply ignored, leaving only the retarded waves that we see. But there have always been scientists who are uncomfortable with this interpretation. You shouldn't pick and choose which aspects of a theory to apply. Either it works or it is wrong. Supporters of advanced waves argue that these waves must be there; we just can't detect them.

Two great American physicists, Richard Feynman and John Wheeler, were the ringleaders in suggesting that advanced waves do exist, and that there was evidence of this in an embarrassing aspect of the way an electron in an atom gives off light. Whenever something glows or appears to reflect light, what happens is that electrons in the object drop in energy level, and each one emits a photon of light with the equivalent energy. When the photons are emitted, the electron appears to recoil, just as a gun does when it fires a shell. Yet there is a problem with this at the level of quantum physics. If you do the math, it seems that because of the peculiarities of the "self-interaction" involved, quantum theory predicts that measurements should become infinite.

As electrons in atoms are emitting photons all the time and we don't have reality collapsing under these infinite values, there must be another explanation for the apparent recoil of the atom. Wheeler and Feynman suggested that there was an advanced wave photon heading toward the electron backward in time, and it was the impact of this photon that caused the "recoil." The mechanism of considering particles to move backward in time is quite common in quantum theory, though no one is sure if this backward time travel is real or just a useful convention.

As yet there is no obvious way of linking the concept of advanced waves and precognition. Based on Feynman and Wheeler's theory it would be possible to use advanced waves to communicate backward in time, but it would involve sending messages in a direction where there was nothing to absorb the photons at the far end. By sometimes putting an absorbing object in the way, a variation in the advanced wave should propagate backward in time. It is a very remote possibility—but the existence of advanced waves does give us some hint that there may be a mechanism, should precognition truly exist.

Because of the mind-twisting nature of precognition with its implications of interfering with the conventional flow of time, it can confuse even those who spend their days thinking about parapsychology. Physicist Jean E. Burns, who takes more than a passing interest in psi matters, comments that precognition experiments have not found any difference between predicting classical random events like a coin toss and quantum random events, where the occurrence is truly random and unpredictable.

"It appears," she says, "that precognition works about as well for quantum randomness as it does for classical randomness. No explanation is known for how this can be." This viewpoint suggests that precognition is not about viewing the future at all—if we could see the future, then the quantum random event would already have occurred, and so would no longer be unpredictable. Rather, Burns seems to see precognition as being able to predict the future from current knowledge, like highly detailed weather forecasting, a process in principle possible in classical physics (though it is often impractical because the systems are too complex to predict outcomes) but not in the quantum world.

The biggest problem with anecdotal evidence for precognition, which is by far the most common way we experience it, is its dependence on our all too faulty memories. We think of a vast number of things in a day. We all experience millions of small, individual occurrences. When an event crops up that we think we may have foreseen, we are selective about what we remember of the circumstances of that prediction, picking out anything suggested by the future event and ignoring aspects that we got wrong. There are lots of things constantly bubbling in our memories, waiting to have some association bring them to light.

When the precognition takes the form of a dream—a common method of attempting to see the future that has been used throughout history—there are even more factors coming into play that make it likely that someone somewhere will have had an appropriate predictive dream. Research suggests that around 80 percent of recalled dreams are negative. Strangely enough, the dramatic events on the news also tend to be negative, making it immediately easier for a dream to forecast a news event without any connection between the two.

Even more interesting, other researchers have shown the perhaps not entirely surprising result that we often dream about subjects that we are worrying about. If you are about to undertake a very dangerous activity, concern about what you are going to do is likely to trigger dreams about things going wrong—which then appear to be precognitive when the likely outcome occurs. In this case the causal link is not a message from the future that comes back to warn you, but rather an expectation of a likely future that then travels forward in time the usual way to meet the actual event.

Although time travel does not contradict any physical laws, the technology that would be required to get a message into the past—a requirement for precognition—is beyond anything we can expect to have in the next thousand years, and there is no obvious means for this to happen naturally. Of course, there have been prophets and sibyls for as long as there has been human culture. We want to know the future and it's not surprising that people have attempted to predict what is going to happen.

Apart from anything else, making a good prediction gives the seer an element of control. Think of the famous scene used frequently in fiction where someone who just happens to know the timing of a solar eclipse is about to be put to death by a superstitious tribe. The person with the knowledge uses his ability to forecast the future to make it seem as if he has control over nature. It would not be surprising if, for example, shamans and priests who became familiar with basic weather lore used that knowledge to give the impression that they could influence the weather and so increase their power in the community. The appearance of being a prophet can be a valuable tool for personal enhancement.

Without doubt the most famous example of large-scale precognition is the writings of Nostradamus. The sixteenth-century French writer, more properly Michel de Nostredame, produced a book of prophecies that, it has been claimed, demonstrated that he foresaw everything from the rise of Hitler and the assassination of John F. Kennedy to the end of the world. However, like many traditional prophecies, the works of Nostradamus are so vague that it is almost impossible to imagine there would be circumstances where he didn't appear to make some predictions.

By coming up with a large number of vague and undated statements, Nostradamus was almost inevitably onto a winner. There has never been a successful prediction of something that has yet to happen from the works of Nostradamus, merely the matching of the verses and events after those events took place—and these linkages are often more tenuous than they first appear to be. So, for instance, the verse usually linked to Hitler contains the word "Hister"—not actually Hitler, but close enough, you might think. However, few Nostradamus interpreters bother to point out that this was simply the old name for the Danube River.

When we think of a more modern approach to precognition than Nostradamus, the kind of thing that springs to mind is the dramatic visions of the "precogs" in the movie *Minority Report.* Even here we don't get absolute certainty—the minority report in the title refers to a dissident view of the future from one of the three precogs. But the movie portrays the gifted individuals getting a detailed vision of a future crime, which is then prevented from happening. The reality of precognition experiments tends to be a search for a much less dramatic ability to predict an emotional response to an image before it is seen—closer to Spider-Man's spider sense than what we'd normally think of as predicting the future.

The clearest indication of a possible precognitive ability from the twentieth century probably comes from Dean Radin's work, appropriately sited for would-be gamblers at the University of Nevada, Las Vegas. Radin and his team showed test subjects images and monitored their physiological response, which, as expected, was different for "calm" images and for emotionally striking images. But the interesting thing was that the subjects' responses started to distinguish between the two types of images during the second or so before an image was revealed on the screen. Radin has shown some selectivity in reporting the success and failure of other experiments, so there may be some doubt about exactly what has been recorded here—but his experiments can be pulled in with some indubitably well-controlled experiments that happened over ten years later as evidence that something is occurring here that needs further study.

Many of the twentieth-century experiments on precognition were sufficiently poorly controlled to be worthless, but there was some surprise in 2011 when respected psychologist

Daryl J. Bem of Cornell University published a paper on what appeared to be genuine precognitive effects. Bem used well-understood psychological experiments, used to study human behavior rather than psi abilities, but gave these experiments a twist that meddled with the timeline, making precognition a requirement for any significant results.

Bem used nine different experiments. When he combined the results from these he produced an outcome suggesting precognition that was sufficiently significant that there appeared to be only a 1 in 134 billion chance of its occurring purely randomly with no cause. This is a dramatic claim. To understand what happened we need to get a better feel for both the experiments themselves and the analysis that produced those impressive numbers.

In one experiment, a group of undergraduates were shown the image of two curtains on a computer screen, one curtain with a picture "behind it," the other covering up a blank wall. The students' task was to click on whichever curtain they felt had the picture behind it. Every now and then, the picture would be an erotic one. As there was no image in reality behind the image of a curtain, just pixels on a screen, this could only be precognition and not some form of clairvoyance, but the experimenters either didn't understand this or wanted to be doubly sure, so the location of the image was not decided until the participants had already made their choice. If the students did better than the expected 50:50 chance of choosing the correct curtain, it seems they were seeing into the future.

The results were fascinating. With the nonerotic pictures, the students did just as you might expect, getting very close to 50 percent right and 50 percent wrong. But with erotic pictures they did better than chance alone predicted, getting the answer right in 53.1 percent of cases. This is not a huge improvement, and with just a few trials it could easily happen by chance. But over the number of tests that were run it was something with only a 1 in 100 probability of happening by chance, without cause. This alone isn't enough to prove anything—for example, when in 2012 it was announced at CERN that the Large Hadron Collider had discovered the Higgs Boson, this was required to have a "five sigma" chance, which is around a 1 in 2 million probability of happening without a cause.

In another test, participants had to choose between one of two pictures that were mirror images, simply choosing the picture that they "liked best." The system then picked, after the students had made their choice, one of the two images. So any deviation from a 50 percent hit might suggest that the participant had somehow got advance knowledge of which image was going to be chosen by the system. Again the statistics threw up a small deviation—in this case, smaller: 51.7 percent were correct, though because there were more trials this result also had close to a 1 in 100 probability of happening by pure chance alone.

Another pair of experiments performed a classic psychological test known as priming, where an individual is asked to decide whether an image is pleasant or unpleasant. Just before, a word, which itself might be positive or negative in implication, is flashed briefly on the screen. When the word lines up with the individual's feeling about the picture, she can respond significantly quicker to the picture than she could otherwise, because she has been "primed." Effectively, her brain is preset in a particular mode, and it then takes longer to switch modes than it does to continue in the same vein.

This is a very valuable psychological tool, as it helps the psychologist assess if someone is being honest. Say, for example, I want to know which brand of chocolate someone likes best. He might actually prefer a cheap, mass-produced, high-sugar brand, but, wanting to look good, he says he likes the fair-trade, high-cocoa-solid brand, because this is considered a more sophisticated choice. His response to priming will give him away, because his response to the image will be based on his actual feelings, not on what he says is the case. What Bem did was to put the priming word up *after* the decision was made, and he still found that it seemed to have a small effect in speeding up the associations that it should have primed for.

Bem also modified a series of experiments dealing with habituation (the reduction of stimulation by repeated exposure to striking images and sensations), boredom, and memory to produce an effect if the participant had foreknowledge of a piece of information, and he found that in all but one of the experiments he undertook there were statistically significant results.

There seems to be something happening here, but we have to be cautious about exactly what. If a participant in, say, the photo-position test managed to identify the position of the photo "behind" the curtain correctly every time, we would know that there was a genuine cause—either she had precognition or she was cheating. But we are dealing with a case of getting 53 percent correct rather than the 50 percent expected by chance. It is only reasonable to suspect that there could be something wrong with the statistical approach used to analyze the data, or some flaw in the experiment's design. These small differences from chance feel more like noise in the system than a true result, and can result from very small errors in the way a trial is operated or analyzed.

One possibility, which Bem has thought through but that many other researchers have not spotted, is that there was a problem with the randomness of the selections. The photo-position test depended on the participant's guessing which of the two curtains the picture would appear behind. This needed to be random, as did the distribution of the erotic photographs among the images. Just as in telepathy tests dependent on a randomly selected subject to be transmitted, it was essential both that the overall sequence used was random and that the subset of the random sequence of values selected for the actual trial did not contain any apparent nonrandom artifacts.

To see why this is the case, let's take a ridiculously bad random number generator that simply always put the picture behind the right-hand curtain. As we saw with the Tart telepathy trainer, human beings are very bad themselves at being random, and will tend to favor one side or the other. If, for example, most people tended to slightly favor the right curtain, then there would be an apparent ability to predict where the picture was.

Of course, no experimenter would have such a bad choice of random locations—but should the selection have had a more subtle pattern, it might have been picked up unconsciously by the participants. After they had made a guess, the curtains opened to show them if they had selected the correct curtain, so they had a chance to pick up on a pattern. (I'm not sure why this was done immediately, for if the curtain had not been opened, the participant would not have been able to pick up on any pattern. It might have been better to operate in blocks of two or more guesses.)

The idea that something as subtle as a variation in an apparently random pattern could be picked up unconsciously by a participant is not just a matter of supposition, and it is something researchers should have been aware of for most of the time that well-structured experiments in psi have been under way. Around seventy years ago a study was done showing subjects a series of symbols. Each time the symbol shown would be either the letter H or the letter V. Some trials were random, but others had too frequent an occurrence of one symbol, or some degree of pattern in the order of appearance of the symbols. Rather than carrying on guessing as they did when there was a truly random selection, the participants started to make guesses that went along with the modification, unconsciously picking up on the pattern or on the increase in frequency of one symbol, and scoring better than they should have.

To try to eliminate this possibility, Bem tried running the trial using a random number generator to do the guessing instead of a test subject. With a random series of guesses at what was behind the curtain, there didn't seem to be any correlation with what was observed with human beings. This seems a rather limited check, though. At the very least there should have been a check to see if always guessing one curtain, or alternating between left and right curtains, produced a better result than random chance predicted. When I asked Dr. Bem about this he pointed out that humans are very bad at making random guesses, but the point is rather that the apparently positive results could be an artifact of the way they were generated, and if always guessing the same curtain, or alternating curtains, produced a positive result, this would be one way to highlight the problem.

It's interesting to consider the response I got to this suggestion. Dr. Bem pointed out that there is a "near-infinite set of guessing strategies" and that the examples I suggested testing, such as always picking the same curtain, are not the strategies that subjects are likely to try. Again, this misses the point and perhaps highlights a difference between the approach of psychologists, especially those involved in the testing of psi effects, and that of physicists. The issue was not to simulate the human being, but rather to attempt to probe the way the random generator matched up with decidedly nonrandom inputs. Of course, there are many possible strategies, and they could not all be tried. But not to try even a few simple possibilities like the ones I suggested seems a little unfortunate.

It would also be sensible to try an artificial intelligence program that could try to spot patterns and to use them to see if that could score better than chance predicted it should. This is not a trivial thing to produce, and may be beyond the programming ability of a psychology department, but though it may seem a heavy-duty test, it demonstrates just how careful you have to be when your results are based on marginal statistical differences from chance. If an AI program started matching better than random chance it would have found some kind of pattern that a human being might unconsciously, at least partially, pick up on.

Of course, you might wonder why there was any probability of the sequence not being truly random. It's understandable why there were problems with producing a random sequence back in Rhine's day (see page 140), when experimenters relied on incredibly poor sources of randomness like shuffling a pack of cards. Any magician will tell you just how easy it is get a card shuffle to do what you want to the cards, and we've all experienced our own attempts at shuffling failing to make much of a change to the sequence of a new deck. Now, though, we have computers.

Anyone who has an Excel program, for example, can use the function RAND(), which according to Excel's help system "returns an evenly distributed random number." Simple. But in practice what Excel has is only a "pseudo random number generator." It uses relatively simple algorithms to generate numbers that do indeed jump around a lot, but aren't truly and unpredictably random.

A basic pseudo random number generator will take a starting value, called a seed, then work out a new value using a formula based on that seed. It will then produce the next value by putting the first calculated value into the formula instead of the seed. It is relatively easy to put together a formula that will produce a sequence that jumps around apparently randomly and that doesn't repeat itself for a long time, but the generator will produce exactly the same results every time with the same seed, and it certainly won't be truly random. Apart from anything else, because the same seed will always produce the same result, the generator can't produce exactly the same value twice in a row, or the result would be to lock it into always producing the same value.

The only perfect mechanism would be to have a device built into the system that measured a physical outcome that was truly random—for example, the timing of emission of particles from a radioactive source. Even if such a source was used to produce a table of "true random numbers," in principle the sequence would cease to be truly random at the point that the table was used, because someone had to decide where to start in the table. Only by building the device into the experiment would there be true randomness at work.

Physical random number generators have been used in a fair number of parapsychology experiments, but few have depended on true quantum randomness, which is definitive and clear-cut, but expensive and until relatively recently technically difficult to monitor. Instead, they have tended to be based on what might be called an almost-random number generator, using a source like electronic noise, which suffers because the original data is not digital like a quantum random decision. This means that there has to be a set of rules for translating the physical input into the digital random series, and this process itself can also be faulty.

Professor Bem also points out that in principle there could be other psi effects at work rather than precognition. If the random numbers are stored somewhere before use, these could be accessed by some form of remote viewing. Or if telekinesis is possible, the participant could be influencing the physical outcome in the test machine, something that would be particularly feasible if the device generated its own random numbers rather than working from a predefined table. The hope was that by using both pseudo random numbers and the output of a genuine random number generator, these effects could be overcome, leaving only precognition as a source, though this confidence seems a little dubious.

Leaving aside experimental errors, the other problem with such marginal data is that interpreting it requires very careful use of the statistics. Bem's calculations have not been criticized, but there are some concerns about the simple statement that a particular outcome is highly unlikely and thus is not true. After all, the chance of the numbers that come up in a major lottery every week being the ones to be drawn are well over a million to one. It's so unlikely that those particular numbers would come up that a physicist would say it could not have happened by chance. Yet a set of numbers is drawn every week. Something has to result from a draw, even though the specific outcome is hugely improbable.

As an aside, one of the best-known examples of stage precognition was the dramatic prediction of the British National

Lottery results on September 9, 2009, by mental magician Derren Brown. Brown's show contained a lot of mumbo jumbo about the wisdom of crowds and getting the value from a panel, but in reality what happened was that Brown appeared to produce a listing of the winning lottery numbers just *before* the numbers were drawn on live TV. Or at least, that's what the viewers thought they saw.

Brown's trick has been much analyzed, but there are two essentials that give hints as to how the trick was worked. One was that the exercise was performed in an empty studio—unlike Brown's usual performances, there was no audience. Second, he claimed that for legal reasons he could not announce the results before they had been broadcast. So although he claimed to have written down the results before the draw, the sequence the audience witnessed was first the live draw, and then, a few seconds later, Brown revealing his results.

This comes down, then, to having a way to get the winning numbers written down between the draw and the revealing, a common enough magician's trick, although the way Brown performed it there would need to be some intervention like a split-screen camera trick to make it appear that he wasn't touching the balls on which the numbers were written between the draw and the revealing. (There is some evidence that such a manipulation did happen, as frame-by-frame analysis shows one of the balls moving slightly between frames.) Brown, incidentally, claims no special abilities and assures us that all his performances are tricks. There was no actual precognition here—and let's face it, if there had been, Brown would have held a multimillion-pound winning ticket in his hand.

If you take a simpler example than lottery draws, coin tossing, and found after 1,000 tosses that you had 527 heads and 473 tails, then this is likely to happen a little less than once in 20 trials—unlikely, but certainly not impossible. However, the slightly mind-bending implication of the statistics is that this *doesn't* mean that it is 20 times more likely that the coin is biased rather than unbiased if this outcome does occur. Say you have a coin that is biased in a way that you expect to come up with 527 heads. Then even with such a biased coin, the chances of getting 527 is only improved by a factor of four—it is still unlikely to happen, because on a single example you can get any one of a range of possible outcomes, not just the most likely value.

There are other potential statistical issues too, and Bem did not run vast quantities of trials. But he did publish enough information for them to be repeated, and to date there has been one series of experiments with similar results and six which did not find any significant data—pulling these together weakens Bem's case. With each test that comes up with no difference from random chance, Bem's positive results become less significant, making them more likely to have been either a rare statistical fluke or a result of experimental error.

Seeing the future is probably the most potentially rewarding psi ability. Just emulate Derren Brown, buy the right ticket, and watch the money roll in. However, there is another favorite of parapsychologists that is probably even more popular than precognition with the military. Knowing what your enemy is up to gives you a huge advantage, whether on the battlefield or searching the caves of Afghanistan for a terrorist. There is nothing the military would like better than someone who could perform the psi feat known as remote viewing or clairvoyance.

6.
SEEING ELSEWHERE

A prim-looking man with a neat goatee beard sits at a table. He wears a suit and his quiet British reserve is quite a contrast to the enthusiasm of his American hosts at Sedona, Arizona. He has asked one of those present, someone he has never met before, to go into another room, down a corridor, and entirely out of sight from the table. The man calls out, asking the woman who left the room to start drawing images.

Over the next few minutes she will draw four items. The man concentrates and calls out what he is seeing. Another of those at the table, a stranger, draws what the British man describes. At the end of the session the drawings are compared. For the first, the man receiving the images has described a religious symbol, like a cross, and a tree or shrub. The original was a Star of David and a tree. The second image he describes as curving, like a banana—which is exactly what it appears to be. For the third he describes being on water and a rocking motion. What was drawn was a boat on water. And for the final image he describes a warm face, like the sun. The image? A smiley sun.

What has happened here could easily be described as remote viewing: the ability to see a remote location without being there. In circumstances like this it is difficult to distinguish remote viewing from telepathy—it's not clear if our expert was seeing the picture directly or seeing into the mind of the person making the drawings. Yet there was no attempt by the person doing the drawing to transmit the information, and the effect either way was to reproduce a drawing from a piece of paper at a distance.

Remote viewing would be a highly desirable capability for military and intelligence agencies, if it were possible. Imagine all the money spent on surveillance, whether it's spy satellites and planes or people on the ground. Think of being able to throw all that away and replace it with a remote viewer who could never be detected, never be captured or shot down. It's no surprise that remote viewing has been of great interest to the military over the years—and were it not for one small problem, the man featured in the event described above would be a hot property in military circles.

The problem is that the man is someone we have already met a couple of times. He is called Derren Brown, and he is arguably the British equivalent of James Randi, an illusionist specializing in tricks of the mind who likes to show how it's possible to reproduce supposedly genuine psi abilities using perfectly natural abilities. The whole demonstration was an act, put on for a TV show. The other participants were totally convinced by the demonstration, and assuming we believe Brown, operating under the strict rules of British TV broadcasting, there was no collusion—those present genuinely believed that they were experiencing an example of remote viewing.

It helped, of course, that those present at the experiment were believers. These were not dispassionate scientists, with no preconceptions about what was and wasn't possible. These were individuals who were already totally convinced that the ability being demonstrated was possible. But even with the knowledge that he was faking, watching the events unfold on video, Derren Brown's demonstration is highly convincing.

Since there is no doubt that Brown's performance was an act—he repeatedly assures his viewers that all his "powers" come down to trickery, psychology, suggestion, and manipulation—it is possible to come up with a number of suggestions for the way that the trick could have been performed. We never see the other room where the individual creating the images was sitting. We don't know, for example, if it had windows through which a confederate of Brown's could have watched, or whether the camera crew had access to the room beforehand and could have set up hidden cameras. In either case, Brown could have had a near-invisible earpiece through which he was fed information, or could have had information relayed to him using visual signals by his camera crew.

Just as telepathy has a parallel in the ways we now use technology to communicate words remotely, the idea of remote viewing using technical aids is a surprisingly old one. We now aren't at all surprised to see TV images from the other side of the world—or even from robotic explorers on Mars—giving us the ability to witness events in some distant location, but even back in the thirteenth century the remarkable English friar Roger Bacon was describing a form of remote viewing using technology.

In his vast book proposal for an encyclopedia of science, Bacon speaks of mysterious and wonderful optical devices. He begins with remote viewing using mirrors:

> Similarly mirrors might be erected on an elevation opposite hostile cities and armies, so that all that was being done by the enemy might be visible. . . . For in this way Julius Caesar, when he wished to subdue England, is said to have erected very large mirrors, in order than he might see in advance from the shore of Gaul the arrangement of the cities and camps of England. Mirrors, moreover, can be so arranged that as many objects as we desire may be visible and all that is in the house or in the street.

In reality the Romans had no such technology, but it gives a good feel for the way that even then it was obvious that the ability to undertake remote viewing would be of huge value to the military. To achieve the effect he describes, some of Bacon's mirrors would have to be curved like those in a telescope to provide magnification, and in reality there would be too much distortion and air scattering to get a good image at this distance even with the best modern technology. Bacon doesn't explicitly mention the use of curved mirrors in this section of his book, though he was aware of their light-collecting abilities and describes them elsewhere. But in his description of hypothetical surveillance technology he does go on to bring in the effects of lenses:

> For we can so shape transparent bodies, and arrange them in such a way with respect to our sight and objects of vision, that the rays will be refracted and bent

> in any direction we desire, and under any angle we wish we shall see the object near or at a distance. Thus from an incredible distance we might read the smallest letters and number grains of dust and sand.

From Bacon's musings we can see a clear path through the development of the telescope, and then to the more indirect mechanism for seeing remotely of television, where an optical image is converted into an electromagnetic signal, transmitted from place to place, and then converted back to the optical to stimulate our eyes. Once we get that intermediate transmission part of the exercise, there is a very little practical limit to our ability to see at a distance.

We are familiar and comfortable with this kind of technology. It's everyday. But is there a way to carry out the same effect remotely without the addition of technology, to see remotely with the mind? To find opportunities for remote viewing to work we need first to understand how we see the world around us. Sight is dependent on light, one of the most fascinating phenomena in the universe. It is often convenient to think of light as being a wave, but the reality seems to be closer to a flow of particles called photons that carry electromagnetic energy.

In the process of sight, a light source like the sun, or an artificial light, first generates a photon, usually as a result of an electron within matter dropping down from a higher to a lower energy level, a bit like an elevator dropping from one floor to the next. If that dropping elevator crashed into something, bringing it to a stop, it would release a burst of energy as sound and heat. Similarly, the dropping electron releases a burst of energy in the form of a photon.

Once created, photons are unable to stand still. Their very

nature requires them to move, as their mix of electricity and magnetism requires a constantly moving electrical impulse generating moving magnetism, which itself generates moving electricity, a process that can only haul itself up by its own bootstraps if it continues to travel at the speed of light.

The photon, then, shoots across space until it hits the object that is going to be seen. Traditionally we would say that at this point in time the object then reflects the photon like a ball bouncing off a wall, but in reality the object absorbs the photon, which kicks one of the electrons in the object up to a higher energy level. Almost immediately, the same electron emits another photon. This is the point at which the concept of an object's color arises. Most substances have a preference for photons of particular energies. White light consists of photons of a whole range of energies, but some of these will be kept by the matter that absorbs them while others knock electrons into unstable states where they easily emit a new photon.

The color of the object we see corresponds to the energies of the photons it tends to re-emit. Different energies correspond to different colors. So if, for example, a particular object tends to hang on to photons with energies corresponding to yellow, green, and blue, but re-emits photons in the red range, we see that object as being red.

At this stage of the light's path, a photon has been emitted by a light source and absorbed by the object being viewed, and a new photon has come shooting out of the object toward your eye. This crosses space again at light speed, a nifty 300,000 kilometers (186,000 miles) per second in the vacuum of space, though it's a little slower in air, rapidly reaching your eye. The chances are that the specific photon will be absorbed by an electron as it passes through the lens and aqueous humor of your eye and then re-emitted, but at the back of your eye one of its successors will reach the retina. This biological projection screen contains around 140 million sensors.

Of those sensors, around 120 million are rods that ignore color and effectively work in black, white, and shades of gray. The remaining sensors, the cones, are split into three types that specialize in the primary colors: red, blue, and green. Each handles a wider range than the primary color alone, providing some overlap, but each of the three sensor types is most sensitive around its particular primary. When the photon reaches the back of the retina (the sensors are back to front, often cited as a good reason for suspecting that eyes are not designed but evolved) it is absorbed by a photoreceptor molecule. As usual, when a photon is absorbed by matter it gives an electron a boost of energy.

The accumulation of energy this way from several photons makes the sensor generate a tiny electrical charge—when the inputs of a range of sensors are combined the result is a signal that is passed up the optic nerve toward the brain. At this point what we're dealing with is electrical signals, and to understand of how vision works it is important to realize that though these early stages bear some resemblance to the mechanism of a video camera, what happens to those signals next is totally different.

In a camera, the signals from the various sensors behind the lens are turned back into colored light. The image that the camera picked up is re-created on a screen by lighting up colored pixels corresponding to the sensors in the camera. The screen shows what the camera detects in a simple correspondence. What it sees is what you get. However, the view of the world provided by your brain is totally different from this. It is a purely artificial construct of your mental processes.

The signals that have passed up the optic nerve are not converted back to an image. What would be the point? There is no little man sitting in your brain waiting to watch the picture. Instead, the brain processes the raw data, using different mental modules that handle light and shade (primarily from those black-and-white rod sensors), edges, shapes, movement, and so forth. The brain then makes up a representation of the world based on the data from those modules. What you "see" is a fiction concocted by your brain.

This construction of an artificial picture of the world by the brain is why optical illusions fool us. We know that what we "see" is artificial, because the image that we think we see is totally different from the image that is being projected on the retina at that point in time. Apart from anything else, part of the retina, where the optic nerves connect to it, doesn't have any receptors. Our brains fill in this blind spot, leaving us unaware of it. And our eyes are constantly making tiny jerking movements called saccades, used to build up a more accurate picture of depth and shapes. If we saw the true picture produced by the eyes it would constantly jerk around, leaving things blurry and unclear. The brain edits out all this noise.

Seeing something, then, is anything but a trivial process. But how could we see remotely? Leaving aside the telepathy-based approach, where the remote photons are falling on someone else's retina and the idea is that a signal then gets to our brain from his, how could we somehow detect the view at a distant location without any optical equipment present to pick up the photons?

If you accept a Descartian split between mind and body (see page 53) it is just possible to imagine that the mind is somehow floating free from the body and travels to a distant location.

This kind of duality is certainly what is imagined by many people who perform remote viewing. They believe that they are having a kind of out-of-body experience, that something inside them—their mind or soul or spirit, depending on their background—has become detached from their physical body and is venturing out to survey the world. It's not at all clear, though, how this insubstantial entity manages to interact with the very physical photons of light to be able to see something. Surely light would pass straight through it? For the photons to be stopped and absorbed there has to be matter present, which wouldn't be the case here.

A useful parallel to think of is the invisible man. Ever since H. G. Wells's novella of that name the idea of being invisible, being able to wander the world causing mischief undetected, has been very popular in fiction. However, a true invisible person would not have nearly as much fun as people suppose—because he would be blind. For the photons of light to trigger the sensors in your eyes they have to be absorbed. But if you are invisible, the photons have to be able to pass straight through you. So they can't be absorbed in your eyes and you can't see. An invisible spirit presence would have the same problem with the added difficulty of getting light to interact with a nonphysical entity. A projected mind would be blind.

If this dualist idea, which is rejected by practically all modern psychologists and neuroscientists, is neither likely nor practical, the alternative is for there to be some kind of link between a human being's mental processes in one location and the photons in a different location. While quantum entanglement can provide for interactions between photons and matter, it has so far always involved the two being collocated initially, even if they are later separated. While it's just about imaginable that two conscious minds could become linked remotely in telepathy, it's hard to see how a brain in one place could somehow tune in to light in another place. For the moment, we probably have to accept that there are real issues with providing anything close to a scientific explanation for remote viewing.

In many of the psi experiments that have been labeled as clairvoyance (the early term for remote viewing) it isn't necessary for something to be visible in order for an observer to claim the ability to detect it. Many such tests, like some variants of telekinesis, are more about the interaction between a mind and an electronic process. This seems to have been the basis of an "ESP teaching machine" devised by physicist Russell Targ and his colleagues at the Stanford Research Institute (SRI) in 1974. Although it had a number of modes that in principle could also test telepathy and precognition, the tests that Targ made concentrated on a kind of man/machine communication that did not depend on any of the conventional senses.

The machine, crude-looking by modern standards, featured four panels, any one of which could light up showing a single image, which was simply a transparency (this was long before LCD panels) and so was fixed during any particular run of the machine. In the clairvoyant mode, an internal randomizer inside the machine chose one of the four panels that would be displayed but did not indicate the selection that had been made. The person being tested (and theoretically trained) for ESP would then choose one of four buttons, one for each of the panels. Finally, the machine lit up the panel that it had already chosen, so the participants could see if they had been successful. A match could be attributed to precognition, seeing the image that would appear in the future, but the experimenters thought it more likely that it was clairvoyance of a decision that had already been made inside the machine.

For the trainer to work, the subject would have to tie in to an electrical signal within the device. Rather than picking up on photons to see an image, the subject would be detecting the electrical charge in a capacitor, or the value of a bit in a section of computer memory. To do this requires two very specific abilities: to detect the presence of that electrical signal or charge, and to clearly identify which is the correct signal corresponding to the specific panel being selected. After all, any electronic device will contain many capacitors or bits of computer memory; it is something of a mystery how the subject was supposed to track down the correct charge in the maze of electronics.

To reinforce the results and provide feedback to encourage learning, the device counted both how many trials were made and how many of those were hits, where the correct image was selected by the subject. If the subject got more and more hits, a series of text panels lit up with sometimes whimsical inspirational messages ranging from "Good beginning" through "Useful at Las Vegas" to "Psychic, medium, oracle."

Initially the scores were recorded manually to test the basic operation of the device and ensure that it functioned at all, but then the trainer was hooked up to a printer that produced a sequence of results rather like the printout on a till roll, recording the position of the attempt within the run of 25 guesses, the value from 1 to 4 that the machine chose, and the number of the image that the subject chose, also displaying a running total of hits. As described this seems a foolproof system, but there is a significant problem with the way that the system was used in practice.

If a subject felt she was not getting any information—if she just didn't receive any impressions of which was the correct image—she could push a button on the machine to pass on a particular guess. When she did this, it wasn't scored. Of itself that shouldn't have been a problem, as the subject could only pass before a guess was made, and the passes were recorded by the printer. Data was not lost. But there was another way to bias things.

In a run of 25 attempts, if the subject was purely guessing with no information received, merely getting as many correct as random chance would predict, she should have got around 6 of her guesses right. But the process would not be random if the subject got 6 on every run. There would be something very suspicious happening if she got the same values every time. Averaged out over many runs, the results should indeed be 6 (or more accurately 6.25) correct answers out of 25. But on any particular run, the result would be expected to be 5, 6, or 7, and would occasionally be even further away from the expected value. Now imagine that the person undertaking the test wanted to cheat. If the subject was responsible for collecting the test results, she could easily do so.

All that would be necessary would be to keep all the printouts from runs that scored more than 6 and to discard a few of the printouts from runs scoring less than 6. Such "cherry-picking" would make it easy to bias the results to suggest an ability that was better than random chance. For this to happen there wouldn't even have to be a deliberate attempt at cheating. For example, the subject could have been disturbed during a run by getting a phone call, or hearing a loud noise, or any one of a hundred different disruptions. If that run happened to have a low score, then the bad performance could be ascribed to the disturbance, and the run discarded.

The same decision would probably not be made if a disturbance happened during a "good" run. Because the results were good, there would be a natural inclination to want to keep them—and clearly if the results were good, the argument would go, there was no need to discount them, because the trial worked. The disturbance didn't make any difference to the subject's performance. If this sounds naive, such cherry-picking resulting in an unintentional bias of the results is found in conventional scientific experiments and probably happens significantly more often than is acknowledged. The archetypal experiment designed to demonstrate this kind of unconscious experimenter bias was a clever trial undertaken in 1963 involving albino rats and, more important, young scientists who believed they were experimenting on the rats, but in truth were themselves the experimental subjects.

Robert Rosenthal and Kermit Fode of Harvard University set up an experiment where twelve psychology students were given five rats each to test on a simple T-shaped maze. All the rats were from the same stock, but one group of the students was told that they had especially bright rats, naturally suited to solving mazes, while the other group was told that their rats were of a less able strain that struggled with maze solving. Those with the "bright" rats were instructed that their animals would show clear learning during the first day of running the maze and after that their performance would improve rapidly. The subjects with the "dull" rats were told that their experimental subjects would provide little evidence of learning.

Both "types" of rat (bear in mind that all the rats were identical in ability) did prove to have performances that improved over time, but every day over the five-day trial the "bright" rats were recorded as performing better than their peers, achieving successful runs up to twice as frequently as the "dull" rats, and getting to a successful conclusion significantly quicker than their supposedly slow counterparts. When dealing with bright rats, experimenters could have encouraged them more, giving them more positive handling, which could have influenced actual performance. However, true improved performances were not necessary, as the experimenters could easily have biased the results, even though they could be totally unaware that they were distorting the data.

One possible approach that would shift the results to match the experimenters' expectations would be if they counted borderline runs as successful with bright rats but not with dull ones. They might also have decided to be selective about which results to record because of some apparently sensible reason (perhaps the rat was distracted by a loud noise), cherry-picking positive results as described above. Inevitably they would not be consistent about how this kind of cherry-picking was applied.

Such bias from scientists looking for the results they expect is not limited to biological experiments. It is thought that Newton's famous experiment in which he "proved" that the rainbow colors are present in white light would not actually have delivered the results he claimed: he saw what he wanted to see to get the results he hoped for. A similar bias seems to have happened on the expedition in 1919 that is said to have confirmed Einstein's crowning glory, the theory of general relativity.

According to general relativity, which describes the workings of gravity as a warp in space and time caused by mass, the sun should bend the light of stars passing close by it, making the light travel closer to the sun than it otherwise would. The effect of this would be to make the stars appear farther away from the sun's disk than they were expected to be. Usually such stars can't be seen, because the light of the sun washes them out, but in 1919 an expedition to observe the total eclipse of the sun on Príncipe Island, off the coast of Equatorial Guinea in West

Africa, was made with the specific intent of measuring the apparent positions of stars when the moon concealed the sun.

The team, led by the great British astrophysicist Sir Arthur Eddington, took photographs that it was proclaimed proved Einstein to be right. For many years it was taken as fact that these results proved the theory. Yet there was experimenter bias that was quite possibly intentional. The facts are that only one plate was properly usable, and while this did seem to confirm Einstein's prediction, results from a second expedition to view the eclipse from Sobral in Brazil were closer to the values predicted by Newtonian theory. Taken as a whole, if all the data from the two expeditions was dispassionately analyzed, the only scientific conclusion was that it wasn't possible to either support or counter general relativity from the measurements taken. Later work demonstrated that this wasn't surprising. In 1962 an attempt was made with significantly better equipment to duplicate Eddington's findings during an eclipse—but even then it wasn't possible to be accurate enough to distinguish between Einstein and Newton.

However, Eddington was in charge in 1919, and he was totally convinced of the correctness of general relativity. He made sure that it was the Príncipe readings that were reported and dismissed those from Sobral. Arguably this was very poor science. As it happens, Eddington's decision was correct. Since then there have been vast numbers of experiments that have confirmed the effects of general relativity in other ways. But at the time, the whole make-or-break decision resulting in headlines splashed across the world's newspapers was made far more on intuition than on an honest interpretation of data.

These examples of serious scientific results taking a biased view aren't given with the intent of suggesting that this is an acceptable thing to do, even if it does occasionally pay off. Rather, they show how easy it is to be biased in the way results are interpreted, producing bad science. However much they dislike the outcome, scientists shouldn't cherry-pick the good results and ignore the data that contradicts their theory. To be effective, all observations have to be published and analyzed.

Although there was a better-than-chance score when the users of the ESP training machine could choose which runs to record and which to discard, when a control was imposed that took away the ability of participants to select and discard runs, the results dropped down to exactly those expected by random chance—and this was despite offering a considerable financial reward for success to one of the participants. The clear deduction at this point was that there was no such clairvoyant ability being demonstrated, but it is worth pointing out that those running the trial had a different interpretation.

Their viewpoint was that the reason results dropped back to random levels as controls were increased was because those controls disturbed the mental abilities. This is a common argument to defend failed trials of psi abilities. It is suggested that controls, distractions, or even the simple presence of a skeptic (particularly a magician skilled in detecting cheats) is enough to put off the psi abilities. But this seems a very unscientific argument. There is no good reason why not being able to cherry-pick the results would stop psi abilities from working. It seems much more likely that a reduced level of errors (whether accidental or intentional) makes the results come closer to an accurate reflection of the real cause and effect.

Perhaps the most dramatic example of extreme remote viewing to be performed as an experiment under conditions that ensured that the individual involved couldn't cheat was an exercise making observations of the planet Jupiter, undertaken under the aegis of the Stanford Research Institute. Two individuals claiming the ability to see remotely, Ingo Swann (of whom more in chapter 8) and Harold Sherman, produced a series of observations on Mercury and Jupiter prior to the first close probes to examine these planets (*Mariner 10* and *Pioneer 10* respectively). This seemed an ideal opportunity to test remote viewing: there could be no question of cheating by taking a look beforehand, as no one knew exactly what the probes would reveal in advance.

Researchers Russell Targ and Harold Puthoff of SRI collected information from the remote-viewing exercise and compared it with the new discoveries that were being uncovered by the probes. The experiment was proclaimed a triumph. *Psychic News* quoted Edgar Mitchell, the lunar module pilot of *Apollo 14* and a great enthusiast for psi experiments, as saying that Swann "described things and gave details which were not known to scientists until the *Mariner 10* and *Pioneer 10* satellites flew by the planets," while Smithsonian astronomer and UFO enthusiast J. Allen Hynek commented, "These are matters which Swann couldn't have guessed about or read." The obvious implication was that there was good evidence here that remote viewing worked in a circumstance where it wasn't possible to cheat, short of the viewers having their own spaceship with which to make observations.

Unfortunately, it is not at all clear that the evidence supports this viewpoint. One of the apparently positive observations made by the SRI team was the number of similarities between Swann's and Sherman's viewings—yet of itself this says nothing about the quality of the information they provided. If they happened to agree on incorrect facts, it does not give any evidence of remote viewing, but rather of collusion. And given that Sherman admitted that the two had met up before the experiments were undertaken, the similarity of their reports is hardly valuable evidence. They could easily have agreed in advance on what to say. What is much more important is how well their observations tallied with the information from the probes—and with the vast amount of even better and more detailed information we have learned since then. I'm going to concentrate on Jupiter, as there was a bigger set of observations of it than of Mercury, although the outcomes were similar for both planets.

In their remote viewings, Swann and Sherman came up with some accurate information. Each identified a handful of true facts—but sadly, these were almost all pieces of information that were well known before the Pioneer probe got there. With statements like Jupiter is a striped planet and has cloud cover and that the sun looks smaller from it (Swann), and that there are many asteroids between Mars and Jupiter, Jupiter bulges in the middle, and it has miles of deep cloud cover (Sherman), there was hardly anything among the accurate facts that couldn't have been gleaned from a quick look at an encyclopedia. The only new information was that Jupiter had rings like Saturn—but this was rather spoiled by suggesting that they were sufficiently similar that what was being seen actually could *be* Saturn, while in fact Jupiter's rings are very different in appearance. It was also said that Jupiter's rings were made of crystals (suggestive of the ice in Saturn's rings), while actually Jupiter's rings have a very different composition, being mostly made of dust.

On the downside, much of what both Swann and Sherman described was simply wrong, often painfully far from the truth and seeming to owe more to pulp science fiction than to real observation. Both spoke as if Jupiter had a solid surface, which it hasn't in any conventional sense. Swann described sand dunes on the surface, mountain chains six miles high, and a high infrared count from the surface, while Sherman told of towering volcanic peaks and a reddish brown crust. Oddly, he even contradicted himself by also saying that the planet Jupiter was a "gaseous mass" (this is also sadly incorrect, even though it is referred to as a "gas giant"). Overall, the majority of statements made about the planet by both remote viewers were incorrect.

There is one really interesting remote-viewing experiment, reported alongside the Swann and Sherman tests by Puthoff when the CIA program it formed a part of was declassified. Remote viewer Pat Price was asked to take a look at a Soviet R&D test facility at Semipalatinsk, given only the detailed map coordinates of the location. Price drew a detailed picture of a gantry crane that is quite similar to a drawing of the site provided by the CIA. However, it is difficult to know how much certainty to give to the description of what Price did, as the same paper describes Swann's remote viewing of Jupiter. All that Puthoff says is, "In that case, much to his chagrin (and ours) he found a ring around Jupiter and wondered if he had remote viewed Saturn by mistake. Our colleagues in astronomy were quite unimpressed as well, until the flyby revealed that an unanticipated ring did exist."

Reading this report from Puthoff you would believe that the remote viewing of Jupiter was a great success—whereas taken as a whole its predictions were disastrously wrong. As we have seen, even the aspect he mentions of Jupiter's ring is misleading. But most important was the omission from the paper of any mention of all the errors that were made. With this level of inaccuracy in Puthoff's paper, it is hard to have any confidence in the assurance he gives that the Price viewing of Semipalatinsk was under double-blind conditions.

What can we conclude about Swann's and Sherman's mental expeditions to Jupiter? Most of their results do not provide good evidence for remote viewing. Leaving aside comments that they made about radiation (how could remote viewers "see" radiation?), there seemed little more than basic textbook research and a lot of very poor guesswork. Swann has since suggested that he missed Jupiter and hit on a similar-looking planet in a different solar system that happened to have these different characteristics. Yet even if he is given the benefit of the doubt for this, the characteristics he and Sherman described were so far off the expectation that it's hard to imagine a planet that would combine all the features described. Planets are either relatively small and rocky, or relatively large and mostly not solid.

A much more likely explanation for the results is that there was no remote viewing in this exercise. It is hard not to conclude that the pair simply made up what they thought would be a good match to the reality of Jupiter, based on poor information. This does not make it impossible that remote viewing exists, but it does mean that we should take any evidence from this source with a significant pinch of salt.

Apart from the Semipalatinsk result, what is described above is primarily what is traditionally described as clairvoyance. However, some experiments would universally be described as remote viewing, and these are the ones that have proved particularly interesting for potential military applications.

In such a remote-viewing test the idea is to be able to look out over a location as if you were there without being present. The standard method for testing for this ability, like Derren

Brown's demonstration at the start of this chapter, seems to accept that there is no physical mechanism for picking up an image directly from thin air, so it relies on the kind of remote viewing that is more like parasitic telepathy, where the viewer attempts to see the world through someone else's eyes, making use of an individual who is already at the location and can provide the mechanism to convert photons of light into a mental message that could in some way be intercepted.

In typical tests, the viewer sits back at the lab with an observer, whose role is to watch the viewer, preventing the viewer from simply secretly tagging along with the away team and spying on those involved. A second experimenter drives off to several remote locations. The viewer describes what he or she "sees" through the remote person's eyes. The final part of the process, sensibly included to avoid bias on the part of the experimenters, is that a (supposedly) impartial judge is taken to the different spots, where he or she tries to match up the impressions of the viewer and the targets, registering how good the match is.

If only remote viewing could be made reliable, as appears was the case with the way the Semipalatinsk test is described, this is a big deal for the military and other government agencies. However, the way the tests have typically been undertaken poses serious problems for intelligence gathering. One is that experiments have rarely come up with remote-viewing exercises where the observation has any significant value. The descriptions tend be so vague and nonspecific that they would be useless for surveillance. To be of any serious use, a remote viewing of vehicles at a location, for example, would need to include the make, model, color, and license plate of each of the vehicles present. What a remote-viewing exercise might come up with is something like "There are a number of lumpy, colorful shapes—maybe cars or houses." This is useless for military purposes and provides no firm evidence.

If remote viewing genuinely involves an ability either to see directly at a remote location, or to see through someone else's eyes, this level of vagueness is hard to understand. Either you see something or you don't. Can you imagine asking someone to describe a scene and them replying with a vague statement like "I see a tall shape and some kind of rectangles. There is an impression of browns and grays," the sort of thing we get from remote-viewing exercises? What is happening here, at best, is remote impression gathering, certainly not remote viewing.

The detailed drawing of the crane in the Semipalatinsk experiment seems exactly what would be needed—yet this does not appear to have led to any significant use of remote viewing in operational military situations. The most likely interpretation is that this is due to unreliability. If Price's Semipalatinsk results were genuine, then they are certainly the exception rather than the rule. Sadly, given the known distortion of the facts on the Jupiter viewing in the paper reporting this experiment, there has to be at least a suspicion that Price may have been given some hints as to what to draw. Standing alone, exciting though the result is, there is nowhere near enough evidence that this was the breakthrough it appears to be.

The other big problem is the quality of many of the experiments both in terms of interpretation of the data and of controls to avoid accidental or intentional bad practice. If you look, for example, at the photographs shown in the book *Mind-Reach* by Puthoff and Targ, you may be particularly impressed by some of the correspondences between the drawing made from the remote viewer's descriptions and the photographs shown.

But as the science writer and skeptic Martin Gardner pointed out, it is entirely possible that these photographs were taken after the descriptions were made and could have been taken in way that maximized the correspondence between the photograph and the description.

What is certainly true is that there is plenty of opportunity for poor deductions from the way these experiments were carried out. All that judges were asked to do was to compare three drawings with three locations and to see if they felt there was an appropriate match. What would have been much more interesting would have been to pick random locations in an area well known to the judge, then to ask the judge which location the image related to. Only then would the drawing be compared with a detailed photograph of the location specified by the judge to see how the details on that day (e.g., what cars were parked there) matched up with the drawing.

In actual examples from Puthoff and Targ's work, the level of vagueness in the descriptions, and the very small samples of possible locations, seem to make it easy to arbitrarily match a description to almost any location. The kind of relatively simple shapes that were described could have been practically anything, and certainly weren't matched to the actual views at the location in any consistent way. An arch shape, for example, could be applied to a rainbow, a bridge, a dome, or any other curved structure. Matching at this level of detail seems worthless.

It is interesting that in their book, Puthoff and Targ describe one remote-viewing test where one of their subjects, the self-professed psychic Hella Hammid, described the target before the target was selected. Presumably the person choosing which target to use had no knowledge of the description given or he could have been influenced in his selection. In this test, three judges all get a 100 percent match between drawing and location. Puthoff and Targ seem to take this as a great example of precognitive viewing—but surely a much simpler explanation is that this demonstrates that there is no remote viewing involved, *because* it worked when the description was made before the target was chosen. There was just matching of very generic images to the desired locations in what was certainly not a double-blind test.

There was certainly a dramatic level of wishful thinking by Puthoff and Targ. Martin Gardner gives a telling example of a target that was actually an arts and crafts plaza. In her description of the impressions she was receiving, the subject referred to a windmill seven times and a golf course five times. The obvious implication from these comments was that Hammid thought that the location was a miniature golf course—but Puthoff and Targ claimed that the match was accurate in almost every detail . . . to the arts and crafts plaza. It is unlikely that this was intentional fraud, but the research involved significant funding from bodies that would only be interested in positive results. A true scientific approach would have considered failure as effective a result as success, but it seems likely that on this particular exercise success was the only acceptable outcome and the researchers made sure this was the case.

We will come back to remote viewing when we look at the military involvement in psi research in chapter 8, but there is one other aspect of remotely sensing the world around us that goes back hundreds and quite possibly thousands of years. This is the form of remote viewing (or at least sensing) known as dowsing.

Dowsing is often treated separately from remote viewing, but in practice there seems to be no good reason for separating dowsing off—the intention is to detect remotely some substance, traditionally water, though more recently the process has also been applied to mineral ores and oil. Before scientific methods for finding underground water, dowsers were in high demand. After all, surely it was better to have some guidance than to simply dig a well randomly. But there was no good evidence of just how successful dowsers really were. Dowsers were (and in some cases still are) hired on the anecdotal reputation of success rather than on any scientific testing of their ability. And inevitably, sometimes you will find water if you dig in what seems to be a likely place, with no need for special powers to provide guidance.

I have tried throughout this book to avoid simply dismissing a particular psi ability as too ridiculous to even contemplate, but I am inclined to make an exception for one version of dowsing. This is where the operation takes place remotely using a map. Instead of passing over the actual materials being detected, where there is some possibility that the operative could detect something physical, the map dowser merely applies his or her dowsing device (usually dowsing rods in the field, though pendulums tend to be used more on maps) to a map of the area of interest and detects water or other materials this way.

The reason I feel justified in dismissing this application of dowsing out of hand is that it is based on a fundamental misunderstanding of what a map is. We are so familiar with maps, and with mentally aligning them with a known territory, that we habitually think of there being some kind of link between a map and the place it portrays. In reality there isn't. A map is just a pattern of ink on a piece of paper, or a collection of data in a computer. It is a representation of a real place, but it has no tie to it—the link between the map and the location is purely imaginary.

There is no way that a sense, however paranormal, can discover something in one by dowsing the other.

Just how ridiculous the idea is can be seen if you look at the process the other way around. If you can dowse on a map and discover something elsewhere in the real world, you should also be able to dowse at the location in the real world and find something that is on the map—which is clearly bizarre. For that matter, let's imagine you know that there is water under a particular spot in the countryside. Then you go to that spot and position a map of somewhere else over the water. Finally, bring in a dowser and ask him to dowse over the map to look for water. How could he or she distinguish between the water actually at the location and something that was being picked up from the map?

Bearing this in mind, dowsing from a map really has no meaning unless the individual knows the location anyway. If he does, a map could be a good way to bring a real location to mind, though photographs should work better—but then, the map is only being used as a prompt to focus on an actual site that the dowser can mentally locate.

Mainstream dowsing is rather different. The dowser typically passes over a portion of land, sweeping it for the detection of water or other substances just as someone would make a sweep these days with a metal detector or a gravitational probe. For the dowser, the detection mechanism is either a pendulum, which will start moving in a particular way when water is passed over, or those most iconic of psychic tools, dowsing rods.

The dowsing rod was traditionally a wooden twig that jerked up or down when detection occurred, though now it more frequently comes in the form of a pair of pieces of bent metal wire, one held loosely by the curled fingers of each hand. The long parts of the wires are held parallel in front of the dowser, and with the slightest movement of the dowser's hands these will twitch wildly and cross each other when passing over something of interest. The traditional twig form has a fork at the end, with one of the smaller sections held in each hand and the longer part of the twig stretching forward. The forked sections are then put under pressure with the palms of the hands, so any small movement results in the twig jerking dramatically, as it is unstable and under tension.

All these dowsing tools have something in common. Pendulum, twig, or metal rods, they are all means of magnifying very small movements of the hands. If dowsing really does work, these movements are being produced by the equivalent of remote viewing. The dowser "sees" the water or other substance and that results in the small involuntary muscle movements that are detected by the rod or pendulum. If dowsing doesn't work, then the apparent detection could still very easily be produced by involuntary movements rather than deliberate cheating—anyone who has ever tried dowsing will know just how easy it is for the pendulum or rods to move with no conscious effort on the part of the person holding them.

The ease with which apparently unforced indications can be generated makes dowsing particularly susceptible to suggestibility. If, for example, a dowser knows that a pipe has water running through it, and walks over that pipe holding her dowsing rods, the chances are that there will be a detection. She doesn't have to cheat (and most dowsers are very genuine about their abilities); she will unconsciously make the tiny movements needed to register the presence of water. This makes it particularly important that a trial for dowsing is done blind, without the dowser knowing where the substance being dowsed for is located. There is no point testing dowsing with targets that are detectable by the other senses.

Dowsing also demonstrates the importance of using *double-blind* testing, because of the Clever Hans effect. Clever Hans was a German horse that became famous in the early years of the twentieth century when he appeared to have remarkable abilities. He could, for example, calculate the answer to simple arithmetic problems and answer other questions that made it seem that Hans could understand German and reason like a human being. After tests by a psychologist in 1907 it was discovered that Hans was picking up on subtle cues from his owner and others around him.

There was no evidence that Hans had been trained to pick up these almost imperceptible cues, as is the case with some other performing animals. Instead, he seems to have matched the small physical movements of his trainer with the rewards he received when he got the answer right. The trainer would inevitably make some kind of movement when the correct answer was reached, and Hans noticed this. You don't have to be a horse to detect these small involuntary movements. Human beings can spot them too. Stage mentalists use them in mind-reading acts (see details of Washington Irving Bishop on page 57) and many of us are able to unconsciously pick up on such prompts.

This means that a good dowsing test must be double-blind. Not only must the dowser be unaware of the location of the target; so must everyone else present from whom the dowser could pick up prompts. This immediately means that the vast majority of apparently controlled tests of dowsing that have taken place to date are useless. While a fair number are blind in the sense that the dowser doesn't know where the target is, very few have been double-blind, with the observers unaware of the location too. Where there has been double-blind testing, dowsers have come out very badly, demonstrating no ability to locate anything over and above blind chance.

An excellent example of really getting the test conditions right was James Randi's 1979 experiment, made with the Italian TV company RAI. At the time, Randi was offering $10,000 (since increased to $1 million—see page 21) for demonstrating a "genuine paranormal feat." Not surprisingly, several famous Italian dowsers came forward to demonstrate their skills in the hope of winning the prize. Unlike the most common dowsing tests, typically involving a series of containers, some containing water and some not, Randi and the TV show team set up a sophisticated experiment with a series of pipes laced in a tangled fashion across a region of ground. Once the pipes were buried, a row of valves could switch the flow along different routes with no visible indication aboveground of which pipes were being used. Neither dowser nor observers would know where the water was running through on a particular test.

Making use of flowing water was particularly useful, as some dowsers claim to be able to detect only water that is on the move. (For some reason many dowsers believe most groundwater travels in underground rivers and streams, and claim to have found these all over the place, though such streams hardly ever turn up in practice. The majority of groundwater is static; in reality there are very few underground streams and rivers.) Randi's setup meant that it could be used to test any dowser, however fussy he or she was about moving water. With the pipe array established in the little town of Formello, around thirty miles outside Rome, Randi was ready to test the four dowsers who came forward. The first came up with a route that ran in a totally different direction from the actual pipe—instead, his suggestion was what you might expect if the water flowed in a straight line from the inlet to the outlet.

The other dowsers fared no better. In fact the second continued to "detect" the flowing water (in the wrong place) after the water supply from a truck dried up. The third claimed there was no water flowing even when he was right over an exposed piece of piping at the start of the inlet flow (it's thought he did this because he couldn't feel any vibration in the pipe with his foot). And the fourth crossed the path in the wrong direction before veering off to the other side of the site. Interestingly, all dowsers were asked if there was natural water on the site. Two said there was none, while the other two described the paths of underground streams—but practically at right angles to each other. There was absolutely no consistency among the dowsers.

Of course these were only four dowsers, a ludicrously small sample, but they were acknowledged experts in their country, the best in the field, and there is no reason to assume that Italian dowsers would be better or worse than those from any other country. I am not aware of a single trial before or since Randi's that has had such an excellent level of controls, and has provided such a good match to a real dowsing environment, that has resulted in a success for the dowser.

One particular peculiarity of dowsing is that some practitioners claim the ability to detect ley lines on the ground or from a map. Ley lines are a fascinating concept, dreamed up by a retired brewer's salesman, Alfred Watkins. Born in 1855, Watkins spent a lot of his working life riding horses from place to place, and over the years he became very familiar with his local Herefordshire countryside. Out on a ride at the age of sixty-five, Watkins was struck by the possibility that there were the remnants of prehistoric tracks dotted across the British countryside, and that these "old straight tracks" could be spotted by the alignment of notches in the hillsides, long-lived trees, and buildings that could have been constructed on such traditional tracks like churches.

Watkins published his idea in the surprisingly readable book *The Old Straight Track* in 1925 and it quickly became famous. Before long, many people were scouring detailed maps, looking for alignments of standing stones, church spires, and other landmarks. Every generation that has discovered Watkins since the 1920s (I remember doing this myself in my teens) has experienced the sudden frisson of delight when discovering such an alignment. Try it and you almost certainly will discover your own ley lines.

For Watkins, ley lines were primitive roads pure and simple, landmarked ways to cross the countryside before the advent of maps and street signs. There was nothing mystical about them, and he had no more idea of dowsing for them than you might dowse to discover the quickest route to the nearest highway. But the romantically named ley lines concept was picked up by the New Age crowd and became linked with dowsing on the assumption that the ancient wanderers who created the ley lines originally, endowed with a now lost wisdom, were plotting and following unknown forces of nature that ran beneath the ground.

Unfortunately, delightful though Watkins's book is (and I would encourage you to dig it out of the library, because it is an enjoyable read), the concept of ley lines is horribly flawed. Of course, there are some intentional alignments, say of monuments built in connection with each other. And there may well have been landmarks set on high ridges to lead walkers toward them. But the fact is that if you take a motley collection of objects like churches and mounds and standing stones, scattered across the countryside, and look for straight lines, you are pretty well bound to find a number of alignments of three places, and could well find some alignments with four or more markers.

When you look at any alignment in detail it is often not particularly accurate—it looks okay on a map but can be many yards away from being a true straight line on the ground. Of course, if the line was just a collection of ancient navigation markers, as Watkins suggested, this isn't too much of an issue. You wouldn't expect prehistoric surveyors to be accurate to the nearest inch. But the more vague you are about exactly pinpointing a location, the easier it is to accumulate more and more alignments. Just how arbitrary ley lines are can be gathered from the way you can take practically anything that occurs frequently across a section of the environment (telephone boxes and cell phone masts, for instance) and find similar alignments.

Ignoring ley lines, dowsing seems a harmless fringe activity (at least harmless unless an oil company pays thousands of dollars for an attempt to dowse for oil deposits) on the edges of remote viewing. All the evidence is that it is purely a matter of self-deception. As to remote viewing as a whole, while there could be a mechanism that is really a variant of telepathy to allow us to see remote places through another person's eyes, pure remote viewing without an observer to make use of seems highly unlikely. And all the experiments to date have produced little if any effective evidence that this is anything more than a mix of self-deception and making things up.

Remote viewing makes up the last of our explorations of the basic components of ESP or psi abilities that could have a physical explanation. So now we are going to change direction and look at some of the best (and worst) examples of attempts to apply science to parapsychology, starting with the first and probably still most famous large-scale systematic study, undertaken in the 1930s by Joseph B. Rhine.

7.
IN THE RHINE LAB

You have been granted a rare opportunity to sit in on a test of psi abilities. You are shown into a spacious but sparsely furnished room with a small wooden table over to one side. The table is about the size of a card table, and, quaintly, it has a patterned tablecloth laid across it so the corners of the square cloth dangle over the edges of the table, as if it were a relic of the 1920s.

At one side of the table a young man, looking uncomfortable in a jacket and tie, is concentrating furiously, his face drawn into a grimace. Facing him, another young man, taller and more authoritarian, watches carefully. Scattered around the table, some stacked on others, are around a dozen packs of cards, while the observer has a record book in front of him, ready to note every action that takes place.

The first man, Hubert Pearce, takes one of the packs of twenty-five cards selected by the observer and shuffles it, after which he hands the pack over to be cut. Then Pearce picks up the entire pack, lifts off the top card keeping it facedown, tells the observer what he believes the card to be, and carefully puts the card on the table, without looking at it. The observer notes what Pearce decided. When the entire pack has been worked through, Pearce picks the pack up, turns it over, and runs through the cards, allowing the observer to note the actual values with scientific care.

After many such tests, the first great methodical attempt to study psi at work, it was claimed that Hubert Pearce definitely had extra sensory perception. The scores he achieved were close to double those expected by random chance—after this number of tests, the probability of this happening without a cause were astronomically small. And despite that cheerful tablecloth, this was no amateur piece of work. We have just witnessed one of the thousands of trials undertaken in a lab at Duke University in the 1930s.

Although there was wide interest in psychic phenomena, and particularly spirit mediums, at the end of the nineteenth and start of the twentieth century, and these often dubious practitioners were examined with mixed results by scientists, many of whom were unprepared for the possibility that the test subjects might cheat, there was relatively little exploration of the mental phenomena that could have a sensible scientific explanation. This was until one man entered the field and had a dramatic effect, so much so that even though he began his work in the 1930s, his is still one of the first names that springs to mind when considering the scientific study of psi. He was Joseph Banks Rhine.

Originally trained as a botanist, Rhine was the one who coined the convenient term "psi," and although there were certainly some severe shortcomings in the way he carried out his work, Rhine was the first to attempt to apply serious laboratory conditions to large-scale trials at the laboratory he set up at Duke University in Durham, North Carolina. Rhine was convinced of the existence of psi abilities and set out to prove their existence under the sort of controls that simply hadn't existed in earlier demonstrations, which rarely strayed out of the "parlor magic" atmosphere of the spirit medium and the séance.

Many people who are skeptical of paranormal abilities in general still have the feeling that Rhine proved the exception, that his work revealed that there really was a small but detectable effect from the influence of the mind. But the actual results from Duke University and the details of the work carried out there are rarely remembered. Only the name remains, with perhaps the image of special packs of cards featuring abstract patterns, as a beacon of this early work.

We need to go back to Rhine the man to better understand his research. His background is often belittled using the dismissive phrase I employed earlier, "originally trained as a botanist," which seems designed to make him sound like little more than a collector of wildflowers who hadn't a clue what he was doing in the big bad world of psychic research. In reality both Joseph Rhine and his wife, Louisa, were established scientists, receiving doctorates in biology from the University of Chicago and with experience working as university lecturers in the subject.

Both Rhines decided that psychology and the paranormal proved a more interesting topic than the biology they were teaching and switched track in a move that was indubitably seen as risky for their careers at the time. Rhine had read of the telepathy experiments carried out by the British physics professor Oliver Lodge at the University of Liverpool, and he seems to have been inspired by this to throw over his career, taking his wife with him.

Rhine was not, as is sometimes suggested, a part-time dabbler, but became as well established in experimental psychology as he had been in biology. (Louisa seems not to have featured much in the actual research, if Rhine's documentation is anything to go by.) They studied for a year at Harvard in 1926–27, where they were advised on, among other things, "the indicia of deception within the field of psychic research" by Dr. Walter Franklin Prince, an Episcopalian minister who worked with Harry Houdini on investigating corrupt spirit mediums. From there, Joseph and Louisa moved on to Duke University to work in the field.

Once at Duke, Joseph Rhine set out on a three-year course of experimentation before drawing his initial conclusions. This was not the end of the trials—in fact, there is still a Rhine Research Center operating today. No longer under the aegis of Duke University, since Rhine retired in 1965, over time it seems to have moved from being a skeptical inquiring organization to being one that promotes and encourages the belief in paranormal activity without that all-important critical eye that accompanied Rhine's original work. So, for instance, although you will find scientific-sounding studies on "Human biofields" and "ESP and motor automatisms," you will also discover a "Psychic Experiencers Group" and other New Age activities at this center (just down the road from Duke University), which claims to be "bridging the gap between science and spirituality."

When Rhine began to look into psi phenomena he started not by trying to observe parapsychology in action or by developing theories for how it could it work, but by trying to classify the different abilities that were represented as ESP. This may seem rather dull, and some scientists might suggest that what we're seeing here is a reversion to the "stamp-collecting" nature of botany (the great physicist Ernest Rutherford famously commented that "all science is either physics or stamp collecting"—considering that biologists in particular spent their time classifying and recording rather than truly explaining). However, Rhine was ready to point out the error other researchers could fall into.

He believed that it was all too easy, should you observe a psi phenomenon, not to be aware of what was actually happening. "We shall see," he commented, "the dangers of experimentally following one hypothesis without full recognition of the other possible hypotheses—perhaps the greatest danger-point in all human thought." A skeptical observer might think that Rhine was referring to the possibilities that people might attempt to fake mental abilities, something he was clearly aware of as a danger from his training with Dr. Prince. But on this occasion he was thinking of the ease with which an observer could become confused about what kind of ability he or she was seeing demonstrated in what was, after all, a somewhat shadowy world.

Take, for instance, the classic telepathy test that Rhine would repeat tens of thousands of times. One individual looks at a series of cards and attempts to communicate mentally what he is looking at to a receiver. If the experiment is performed with the correct controls there should be no way for the receiver to know what is on the card using her ordinary senses (hence "extra sensory perception" or ESP, a tag that, like "psi," Rhine coined). But even if such a test was 100 percent successful it would not establish the existence of telepathy.

Imagine instead that the receiver was able to perform remote viewing. In this case, rather than receiving a thought from the sender, she would be looking at the card, either directly or through the sender's eyes. But the outcome would be the same as if there had been telepathy. As Rhine pointed out, the kind of test that he often used "dealt with undifferentiated E.S.P.; either telepathy or clairvoyance or both." In fact, Rhine suggested, practically all experimental tests for telepathy up to that date, including his early work, failed to exclude the possibilities of clairvoyance or remote viewing appearing to be telepathy.

Once you start to look at the problem through this viewpoint it becomes quite difficult to devise an appropriate test to isolate telepathy, depending on just what psi capabilities you allow for. If you are trying to exclude the possibility of remote viewing, then the sender would have to see the card when the experiment was not under way and merely think of it during the experiment. That way, the receiver can't be peeking at the card using psi powers. But what this does not rule out is precognition. If it were truly possible for the receiver to see into the future (as we have seen, this seems an unlikely psi ability) then it would be very difficult to totally rule this out as a means of discovering the information, as inevitably at some point in the future someone must discover what the actual values were. If you can see anywhere in time and space there is very little that can be kept secret from you.

To avoid precognition, the results would have to be compiled double-blind in such a way that the person recording the results did not know which receiver he or she was dealing with. The only get-around left at that point is that the receiver had some form of postcognition or retrocognition—the ability not so much to see into the future as into the past. At some point the sender would have to be told which card to send (unless the sender decided off the top of his or her head which card to send, which we will see brings its own difficulties), and at that point the receiver could also be looking over his or her shoulder from what would be the future at that point in time.

It's quite surprising that precognition seems to have been searched for a lot more often than postcognition. Looking into the future seems a much more unnatural ability than looking into the past. It just happens to be more exciting—after all, there is little advantage to be gained from "predicting" what last week's lottery results were—and precognition is easier to test, because there aren't any other means by which the information could get to the subject. Of course, it would be useless to attempt postcognition applied to a generally known fact from history. But attempting postcognition for the values of cards selected a week previously wouldn't be useless, and it is quite surprising that more work has not been done on this.

One way to get around the possibility of postcognition happening rather than telepathy would be if the sender never looked at the card but instead was fed the details of the card through earphones. Leaving aside the possibility that if there is remote viewing it might also be possible for there to be remote hearing (again something we hear very little about), this would at least prevent any form of visual peeking, past or present—unless it was possible to look at the source where the data was being fed through to the earphones.

Rhine started by examining the existing data from trials that had been undertaken over the previous fifty years. These had very varied approaches and (often negligible) controls, and it proved difficult to pull together an overall picture, though the general feeling he got seemed to be that there was a degree of success, but one that depended on relatively small but statistically significant variations from what would be expected by random chance.

It's worth briefly exploring what is meant by "statistically significant," as the term is bandied around a lot in science and it is hugely important in lab-based ESP research, which often relies on statistical evidence. With the techniques used to spot telepathy and other psi abilities in action there will usually be a possibility of getting an answer right by chance. Let's imagine you took an ordinary pack of cards without a joker, shuffled them well, and took out one at random, then asked me, without my seeing it, to tell you what card it was (ignoring the suit). My guess would have a 1 in 13 chance of being correct even if I had no knowledge or psi ability whatsoever.

To demonstrate that there is a psi ability at work, I would need to do significantly better than random chance would allow. If I got the card correct every time it would be pretty obvious that I was getting the information somehow—whether by ESP or otherwise—and that would be the ideal, though bizarrely, most psi researchers seem to consider a perfect score suspicious because it is "too good." But in practice a score that is considered a success will usually be one that falls in a range somewhere between random chance and perfection. The question is, where to draw the line. You might think anything better than 1 in 13 was a positive result, but it's not that simple.

When we say I have a 1 in 13 chance of guessing the card right, we don't mean that I will get every thirteenth card right—in fact, it would be decidedly spooky if I did. To get a better feel for the impact of probability, let's do a real experiment looking at a simpler challenge: guessing whether a toss of a coin will come up heads or tails. Assuming it's a regular, fair coin there is a

50:50 chance of getting it right. Each possibility is equally likely. So if I make a prediction that the result is going to be a head, I have a 1 in 2 chance of being correct with no foreknowledge. I ought to get around half of my guesses right.

A few moments before writing this paragraph I performed a very crude precognition experiment. I predicted ten results of tossing a coin, writing down my predictions, and then I performed the ten coin tosses to see how well I did. (Why not have a go at this for yourself? There is nothing like trying out something to get a better feel for it.) This is a genuine experiment, which I actually performed, with no cheating, though because I did it while alone, it would not be considered acceptable as scientific evidence because it was not controlled in any way. My prediction was:

HTHHTHTTTH

where H is a head and T a tail. Next I performed the actual series of coin tosses and got these results:

HTHHHTHHTT

I have to confess, when I got the first four values in a row correct, I began to get a little nervous. Note, by the way, a not uncommon occurrence of real randomness that we would be unlikely to predict if we tried to write down a random series of results. Six out of the first eight throws were heads. This doesn't show that the coin was biased in any way. A random sequence does tend to have longer runs of repeated values than we would predict. Bear in mind that the coin has no memory. After it has already produced five heads in seven tosses, the chance that the next toss will be a head is still exactly 50:50.

Now let's take a look at how my scoring went in terms of the hits I had made after a certain number of tosses. We'll put the two sets of values together:

Prediction:	H T H H T H T T T H
Actuals:	H T H H H T H H T T
Hit?:	Y Y Y Y N N N N Y N

So after the first four throws I had a 100 percent, perfect hit rate. If I wanted to cheat, that was the point to be selective and stop the experiment, proclaiming myself a master psychic. I then had a run of bad luck. After five tosses I had 80 percent correct, after six it was down to 67 percent, after seven it had reached 57 percent, and on the eighth throw it evened out at 50 percent. Just what you'd expect from pure chance.

Now here's one of the interesting artifacts that emerge from probability and that can catch out the unwary who try to use probability to decide whether or not there are any psi effects. Whatever the outcome, the next throw had to move me away from that perfect random chance result. With an odd number of trials, I *have* to score either higher than chance predicts or lower than chance. (This is particularly important because some parapsychologists claim that getting a guess wrong more often than chance, an effect called psi missing, is just as much an indication of ESP as getting it right. So whatever happened on that ninth throw—or any other odd-numbered throw—they would claim it was evidence of psi ability, had the run ended there.)

As it happens I managed a hit on the ninth toss, so my success rate crept back up to 56 percent, before sinking back to 50 percent with the final coin. This was, to be honest, a fluke—it could with exactly the same probability have gone up to 60 percent if that last toss had gone the other way. My sample was too small to have any confidence that I would get close to 50 percent correct.

This is a very neat little experiment to demonstrate just how careful you have to be with reporting on experiments where the outcome is dependent on statistics. The first trap is cherry-picking. This particular experimental run, genuinely undertaken, demonstrates perfectly the possibilities of cherry-picking. If I decided to use only the first four items of the run, I had a perfect score. If, instead, I felt there was a reason to pick out the second group of four, I had perfect failure. And if I take the run as a whole I have just what probability predicts for random guessing: 50 percent correct. Of course, these samples are much smaller than in a real psi experiment—but we have to remember that many examples of psi "success" are dependent on as little as the equivalent of guessing right 52 or 53 percent of the time.

What this massive variation of possible outcomes by being selective shows is that it is essential to decide in advance how many tests there are to be in a run and then to stick to this, not to find an excuse to drop certain segments of the data or to stop too early. (I dropped the coin partway through my test, for example. If I had decided to omit the next few guesses because of this, I would have hopelessly skewed the results.) As described here, making such a selection sounds like an impossibly silly thing to do. It's obvious, surely, that by extracting either the first four or the second four results and treating them in isolation we are removing any validity from the test. But even scientists can fall for this temptation.

To see how this can happen, consider a slightly more realistic experimental setup. A person is to be tested for his psi abilities—perhaps he is performing telepathically or using remote viewing to detect the values of a series of cards, as we saw in Rhine's experiment at the opening of the chapter. Such tests are almost always done in limited run lengths—25 guesses at a time with the card packs Rhine used.

Let's imagine we do a first run, in effect a warm-up to get used to the experimental protocols. And it goes badly. It seems reasonable that the subject or the researcher might decide to discard these results. It was, after all, just a warm-up. It wasn't intended to be recorded. But imagine that the first run went really well with, say, 20 hits. Then surely would come the justification of including it in the recorded results. It would be a shame to waste such an excellent run. Again the reasoning seems fair. So you count it in. And without necessarily even realizing it, you have indulged in disastrous cherry-picking.

The second observation to make from my little experiment is that I need a big enough sample before I can be reasonably sure that what I see reflects the actual outcome, because random variation will mean that over relatively small samples (the first four throws, for instance) there will be no resemblance to a meaningful run. Exactly how many trials we have to make depends on the nature of the test and what is being compared. This is where the aspect of being "statistically significant" comes in, and we'll come back to that in a moment.

I'd also note that the outcome of any test is highly dependent on exactly when you decide to stop. Note, as I discovered above, that a run like this that we expect to come out with 50 percent correct guesses will *never* produce that exact value if there are an odd number of tests in the run. With an even number of test I could get half hits and half misses, but with an odd number I have to, at the very least, have one more hit or one more miss than 50:50. Even as innocuous a factor as whether the number of tests is odd or even can influence the outcome.

There is actually (at least) one more feature of my trial to consider, arising from the physics of my chosen test, coin tossing. This is the fact that coin tosses depend slightly on the initial state of a coin. Because of the way we flick coins, there is a slightly greater chance of the face that is up before we flip the coin also facing up after the toss. It's a small probability. If, for example, you start with the head up there is around a 51:49 chance of getting a head as a result. But this does mean that the starting position of the coin ought to be evened out, for example by ensuring that the coin tosser alternates between heads and tails facing upward at the start of the test.

You might think that with these simple checks on the validity of your approach under your belt, you are proof against any problems with analyzing a statistical outcome, but as many scientists without a good training in statistics have found to their cost, the devil of stats is in the detail. Try this little coin-flipping puzzle and see if you would come to the right decision and can interpret the outcome of the experiment correctly.

Imagine that to do my precognition testing, rather than flipping the same coin over and over again, I have a huge stack of coins and flip each of them once, one after another. These are fair coins, with a fifty per cent chance of coming up heads or tails. I flip the coins one after another (leaving the flipped coins on the table) until the sequence H T H comes up. At that point

I stop and count the coins. Then I repeat this experiment many times.

For the second part of my experiment I again flip the coins, leaving them on the table, until the sequence H T T comes up. At that point I stop and count the coins. Then I repeat this part of the experiment many times. On average would you expect it to take more flips to produce a sequence ending in H T H or more flips to produce H T T, or would each require the same number of flips?

Common sense says that there is a pretty obvious outcome. Unless you are using psi to influence the flips, it is clear that it will take on average the same number of flips to get H T T as to get H T H. And certainly if I take three coins and flip them, there's the same chance of H T H or H T T coming up. But, remarkably, things are different in the experiment above. On average it will take fewer flips to produce H T T than it would to produce H T H.

Just take a moment to think how that might be possible.

Here's the sneaky probabilistic component that isn't obvious and that many scientists would miss. In both cases, you need the values H T to come up first to achieve the sequence that means you can stop. Now imagine that you then get the wrong face on the next flip. So if you are looking for H T H you actually get H T T and vice versa. Now H T T has an advantage. If you are looking for H T T and actually get H T H, then the last coin in the failed sequence is H. So you now only need T T, just two flips, to complete your sequence from here. If you are looking for H T H and actually get H T T, then the last coin in the sequence is T, so you need all three of H T H, at least three more flips, to complete the sequence.

The reason H T T does better is that the sequence of faces

that *isn't* correct ends in the face that starts your sequence. If you are aiming for H T H, getting the wrong result produces a bad starting point, so you have to run the exercise a little bit longer. This little quirk of probability would be enough to enable you to get better odds than you might expect in a precognition test. Or you could run it as a telekinesis test, trying to force the sequence H T T to appear—and you would score better than trying randomly to produce H T H. Admittedly it would be an unusually convoluted test to make, but it is conceivable that just such a trap could result in apparent psi emerging purely from the statistics.

As we've already seen, it isn't enough just to have, say, 51 correct guesses where 50 is expected. The experimenter has to have enough extra correct guesses to be reasonably confident that the result couldn't have happened by chance, and this is where the concept of statistical significance comes in. It's a measure of how likely something is to have happened by chance, and hence of how much confidence there can be in there being a cause for the event.

This can be worked out because random occurrences usually fall in a distribution. Though not universally applicable, the best-known such distribution is the normal distribution, or "bell curve" as it is sometimes known because the shape of a plot of the values in such a distribution looks a little like a bell. It has a high peak, centered on the expected value, then drops away to both sides, ending in long tails with increasingly low occurrences, but which are still expected occasionally. So in a set of coin tosses, the more repeated runs you make, the closer you would expect a plot of the possible outcomes to match a normal distribution.

Once you know the distribution that applies, it is possible to assign a "significance level" to whatever result is actually achieved. So, for instance, a 5 percent significance level would mean that in 5 percent of cases the results of the experiment would be reached purely by chance, but you have 95 percent confidence that there is a cause. Although a 5 percent level is often adopted in the human sciences as sufficient for proof, most other sciences consider this far too high a chance of getting it wrong and are likely to go for 1 percent, 0.5 percent, or better. If there are no other factors influencing an outcome, it is possible with relatively simple mathematics to say how many "hits" need to be achieved in a run of tests to get any particular significance level.

The impact can vary hugely depending on the size of the sample. Take my simple experiment on coin tosses and imagine I flipped the coin a total of 100 times. We would expect by chance that I would get 50 guesses right and 50 wrong, but what if I got 60 right? How significant would that be? The significance level in this case is around 2.8 percent. It's relatively unlikely, but it's going to happen around 1 in every 35 times the experiment is run, so it certainly isn't conclusive. You wouldn't risk your life on it.

Let's say we continue with the experiment until I have made 1,000 predictions and I still get 60 percent right. A gut feel might be this should be 10 times better as a result, so instead of it happening by chance 1 in 35 times it would be 1 in 350 times—unlikely but still entirely possible. In fact, though, by the time we've made 1,000 predictions the chances of getting 600 of them right by chance is less than 1 in 7 billion—significantly worse odds than those of winning a big lottery. Under such circumstances you can break out the champagne. And this reflects the importance of the work Rhine was to do. He made enough tests that, subject to there being no other potential causes like cheating or selection of data, it was highly likely that psi was involved. Many of the tests that Rhine performed had results that were billions to one against occurring by chance alone.

There's a significant pitfall to watch out for here. Some observers who should know better have taken these huge odds as proof of the existence of telepathy or other psi abilities. After seeing dramatic statistics from a particular series of experiments, Charlie Broad, then professor of philosophy at Cambridge University, commented that the results were "statistically overwhelming for the occurrence not only of telepathy, but of precognition." As a professor of philosophy, Broad should have realized that by showing that something is *not* attributable to chance we don't prove that it has a particular cause, merely that it has *some* cause. These statistics and others, including those from the vast number of trials undertaken by Joseph Rhine, do demonstrate that the results are not due to chance alone. They *don't* tell us whether the outcome is due to psi abilities, mishandling the data, or deception.

One key observation, Rhine felt, was that some individuals performed significantly better than others, and that if you used only the data for these individuals, then the result would be to give an incredibly high level of confidence that there was something happening over and above random chance. "Now, this selection is permissible, of course, only if there is valid reason to suppose individuals may differ in their 'guessing' ability," Rhine observed. But there is a huge danger in taking this stance that is reflected very often when those involved in nonrigorous psi trials report their results.

What is true is that it would be perfectly feasible to select those individuals who appear to do well in a first set of trials and retest those individuals, discarding the initial results. That way, if they scored well because of some innate ability, they should be able to continue to score well in subsequent trials. However, to select only the scores of the people who did well from the original trials and to use those as actual results would be totally wrong—it is cherry-picking of the worst kind because it boosts the scores without any actual improvement in results. In any group of values, even those that are totally random, some will be higher than average, some lower. To pick the "best" performers as evidence of a special ability is meaningless.

Imagine that there is no actual effect being measured, and the number of "hits" predicted to occur by chance is 5 out of 25 attempts. In practice, you would not get 5 answers each time the experiment was run. Five would be the mostly likely outcome but you would get regular 4s and 6s, occasional 3s and 7s, and just occasionally you would get results outside even these bounds—but over a long run of trials you would expect an average of around 5.

However, should you decide that one particular group comprises good telepaths as a result of their scores and count only those scores, you have immediately pushed the average up far above 5. The actual statistics only show the expectations of chance, but with this cherry-picking you can make the outcome appear to be highly significant. There is no doubt that such an approach, justified by the thought that it merely involved picking the "best" people, has been responsible for the apparent successes in some ESP tests. Unfortunately, it's easy enough to do this and simply discard the other results without ever reporting that they exist. Instant triumph.

To be fair to Rhine, he was not proposing taking this cherry-picking approach as a matter of course. He made the comment when he was assessing previous experiments before going into his own, and one of those whose work he commented on was a Professor John E. Coover at Stanford. Rhine pointed out that while Coover's results were only mildly positive (with a roughly 1 in 40 possibility of occurring by chance) and Coover considered them to be insignificant, there were a number of individuals who did much better than the others. Rhine pointed out that if you considered only the results of these individuals, then the results would "become tremendously significant—20 times the probable error." Rhine's main aim in making this selection was to point out that Coover should have retested just these individuals—entirely acceptable provided he didn't carry forward the cherry-picked results. But his statement about the selection being permissible is highly dubious.

This tendency to be careless with the numbers seems to have plagued early parapsychological research, which is one reason that the claims for results of many of the early trials have to be considered at best doubtful. Psychology in general as a branch of science has more of such problems than most. As recently as June 2012, for example, Belgian psychologist Dirk Smeesters working at Erasmus University in Rotterdam, the Netherlands, resigned because a panel found that it had no confidence in his scientific integrity, based on his cherry-picking activities.

The aspect of Smeesters's work the panel objected to was massaging data in some of his papers to strengthen the desired outcomes. An unidentified whistleblower examined Smeesters's results and decided that they were too good to be true. The technique Smeesters used was to remove from his analysis any data from participants where they didn't seem to have read the study instructions carefully, if, and only if, removing that data made the overall results more favorable for his desired outcome. Smeesters claimed that this was not unusual in his area of work. The culture, he argued, was such that "many consciously leave out data to reach significance without saying so." This is inevitably a major concern when considering the robustness of the historical data of psi testing.

Rhine began with some simple trials to see if he could identify some strong subjects for more detailed testing. It is interesting that taken together these early tests totally failed to come up with evidence of psi ability, yet this seems to be generally ignored when considering Rhine's contribution to the field. It is an easy mistake to make to consider negative results a failure—and even the best of scientists may be disappointed by a negative outcome. Yet the fact that the outcome is negative is valuable evidence in itself. It is *not* a failure, but equally important to any positive outcomes.

In the first trials, Rhine ran "guessing contests" among groups of children at summer camps. (It's hard to imagine a psychologist being able to do this now without parents signing a whole pile of disclaimers.) Rhine held a card with a number between 0 and 9 stamped on it, looked at it, and concentrated on it. The children had pencils and cards and wrote down what they thought the number was. After around 1,000 trials the result was that none of the children stood out as having clear psi abilities (positive results could have been due to either telepathy or remote viewing).

In a second trial, aimed more specifically at remote viewing, similar numbers (and sometimes letters) were sealed in envelopes, which were passed out to students in classes run by Rhine and his colleague Karl Zener. The students were asked to contemplate the envelope "under certain conditions of quiet and relaxation." After around 1,600 attempts the experiment was given up "partly because it was quite laborious and partly because of the indications of failure." Overall there were not significant results, though one subject was identified as worthy of further study, as he was the highest scorer in both runs of the trial he took part in.

Although we don't have much detail, it does seem that both these trials were open to dangers of cheating, though as it happened there is no evidence that any happened. (It's hard to imagine anyone would cheat in order to make the result more like those expected by chance.) In the test with the children, the card Rhine was looking at seems to have been held openly, with the potential for sneaking a peek, while in the student version we aren't told if any precautions were taken to avoid the students attempting to discover what was in the envelope.

Rhine does, however, make one comment about his experimental procedures that seems to indicate that there was a significant danger of his being taken for a ride by those involved. He says:

> In the conduct of these experiments there has not been a carefully drawn up plan of procedure right from the start. In work of this kind it is necessary to proceed as explorers, ready to adjust plans at every turn, flexible as to methods and conditions. Only the general objectives need to be kept fixed, and the means and criteria of interpretation. Often a block of work is of little value because of poor conditions of security against possible errors or deception, but if thereby there may be a chance to develop a good subject for

> later improved conditions, we relax the conditions and record them as they actually are.

I am no magician (despite a brief dalliance with my high school's magic society) but I do know a little bit about controlled experiments, and there seem to be several points that are worrying in this frank confession, even if we should congratulate Rhine for being so honest. While it may be necessary to revise protocols if they prove insufficient to the task, every effort should go into making the controls effective from day one.

Not only is the flexibility Rhine describes dangerous, in that it can be easy to ascribe results to conditions that didn't actually apply; there is also the problem that it is almost impossible to ignore any positive results coming out of the badly controlled trials. To quote Rhine himself, referring to such a trial, "Mr. Lecrone's conditions were not perfect but they served after 1,710 trials to convince him of the reality of extra-sensory perception." Despite admitting to poor controls by A. E. Lecrone, a skeptical former student of Rhine's ("not perfect" could, of course, cover a multitude of sins), these trials convinced Mr. Lecrone and were considered worth reporting by Rhine. The damage done by poor controls is excellently demonstrated here.

It really is hard to understand why Rhine should need to relax "conditions of security against possible errors or deception" to be able to "develop a good subject." It sounds awfully like making it easy to cheat to begin with until a subject is familiar with the experimental procedure and so can use more sophisticated cheating. Once good controls have been developed, why relax them at any point unless it is under pressure from someone who wants to bypass them?

The subject deemed worthy of further study in the early trial was a Mr. A. J. Linzmayer, and after a couple of trials with borderline results, the Duke team was able to get Linzmayer back for more detailed testing. Joseph Rhine rather neatly sums up Mr. Linzmayer:

> [He] was an undergraduate in this University when he began to work with us. He is of German-American stock, has excellent health, and is a normal, alert and intelligent young man. He is fairly sociable and makes friends easily. Although he is very dependable and even somewhat methodical, there seems to be dash of the artistic, too, in Linzmayer, pretty much undeveloped. . . . There has not been the slightest indication of dishonesty in Linzmayer. He has been scrupulously careful to avoid having any undue advantage given him.

By now, Rhine's handwritten cards had been discarded and his team were using the standard ESP or "Zener" cards, designed for Rhine by Karl Zener. Each pack of 25 cards had 5 each of a set of very distinct designs: a circle, a cross, wavy lines, a square, and a star. These shapes (which are intentionally made up of between one and five lines in the order above) were chosen in preference to conventional playing cards because they are much more distinctive as mental images. The hope was also to avoid a suggestion that some people would prefer particular cards. (This isn't an effect that the Zener cards totally deal with. In the way that many people will choose 7 when asked to pick a number between 1 and 10, many people have preferences for some abstract shapes over others.)

Linzmayer proved remarkably successful. In 600 attempts, for which random chance would predict around 120 correct guesses, he got the correct card 238 times. This far exceeds the possibilities of pure chance—it's a literal one-in-a-million probability: not an impossible outcome without a cause, but very, very unlikely. Rhine points out that in one series of 25 tests, Linzmayer achieved 21 hits, 15 of them in succession. This remarkably successful run took place in the rather unlikely laboratory of Rhine's car, with the engine going, presumably to provide a background noise to reduce the chances of involuntary whispering being heard.

According to Rhine, Linzmayer was leaning back over the seat (this sounds highly uncomfortable) so that he was looking at the roof of the car. Rhine was sitting next to him, drawing cards from a pack, taking a peek at the value and placing them on a record book lying on Linzmayer's knees. The chances are high that Linzmayer either had a genuine talent or was cheating (consciously or unconsciously), something that unfortunately was all too possible with the early Zener cards, and particularly in these bizarrely loose conditions in the car.

Even Rhine accepted that there were issues with the way this trial was conducted. He noted that Linzmayer called out what the card was and Rhine said "Right" or "Wrong." This in itself was unfortunate. Given that there are only 25 cards in a Zener pack, knowing what has gone before can easily influence later guessing. Card counting is much easier with a single Zener pack than with a blackjack shoe in a casino—and is something it would be entirely possible to do without being consciously aware of it.

What's more, Rhine tells us that whereas ordinarily he would record the results after every call, this time he continued for most of the cards without a break, remembering the outcomes and noting them down only at the end. "The easy informality of this situation may have made the brilliant run of 15 unbroken hits possible," he comments. This is painfully true, even if not necessarily as Rhine intended it.

In describing this run of 15 hits, Rhine falls hook, line, and sinker for the cherry-picking approach to statistics. "The probability of getting 15 straight successes on these cards is $(1/5)^{15}$ which is one over 30 billion," he comments. While there is no doubt it is pretty unlikely it is still a very small sample, selected because of the way it stands out. Remember my coin-flipping experience. If I extract the first four tosses, all hits, I get a very different picture than I do from the whole run. As soon as a selection is made from the data based on the values of those selected points, any attempt at a scientific view goes out of the window.

In retrospect it seems clear that there were major issues with many of the tests using the Zener cards, which would have been exacerbated if the participants had an incentive to cheat, like a financial reward. The original cards were printed on paper that was sufficiently thin that it was possible to see the design through the back of the card. They were also originally poorly constructed, with the cards for some symbols larger than others. Later, when the cards were made thicker and opaque, there were still problems with some symbols being readable from the back because the different symbols produced different markings on the edges of the card. But even if the card design had been perfect, it would still have been possible to cheat because the sender and the receiver were in sight of each other, often sitting face-to-face across a desk.

There are well-known techniques employed by magicians to cheat in these circumstances. It is sometimes possible to see the card being looked at by using a reflection, for example, in the sender's glasses or in a piece of metal in the room. It's sometimes even possible to use the pupils of the sender's eyes as a mirror. In these circumstances, the simplicity and distinct nature of the five images in the pack made them ideally suited for a cheat. The slightest glance of even a distorted reflection of a star or wavy lines, for example, would be enough to identify a card.

Similarly, if the receiver was ever left alone with the cards it would be easy enough to rig them so that they could be read from the back. This could be done, for example, by making small nicks and creases in the cards, or making tiny pencil marks on the back. Of itself, this opportunity to cheat doesn't mean that a particular experiment *was* corrupted, but without checks to ensure that the cards could not have been marked, little can be read into apparently successful results. In the early years of such research in particular there were very limited controls and poor recording of just what was and wasn't kept secure by the experimenters.

Finally, there is the huge problem of randomness. Manual shuffling is a very unreliable way of randomizing a set of cards, and in many of the Rhine trials, the subject was allowed to shuffle his or her own cards. If done properly by an expert a shuffle can be reasonably effective, but there are ways of shuffling that will result in a very specific ordering of the cards if the shuffler has appropriate skills, and at the other extreme, many people are simply bad at shuffling. Depending on how the Zener cards were arranged in a fresh pack it is entirely possible that they ended up in a partially predictable state after the shuffle.

Even if the shuffling was perfect, the very nature of a Zener card deck—or at least the way it was used—means that there would inevitably be deviations from random expectation. The problem is that the deck contains only five of each symbol. Because the cards were discarded after each guess rather than reinserted into the pack, this makes runs of a symbol less likely than would happen with a true random sequence. If the first two cards I turn up from a Zener pack are both triangles, then for true randomness there should still be a one in five chance that the next card is also a triangle. The selection should not have a memory of what came before.

In practice, though, I am left holding a pack with three triangles and five of each other card. So the chances of the next card being a triangle is 3 in 23 (0.13) rather than 1 in 5 (0.2)—already significantly reduced. Combine this automatic tendency of the pack to produce fewer long runs than randomness requires with the human being's natural tendency to dream up sequences with little or no repetition and you will get a totally fictitious apparent psi ability that emerges from the statistics alone. In fact if you match one deck of cards as "guesser" against another deck of cards as "sender," the guessing cards themselves appear to have a mathematically significant psi ability because of this statistical anomaly. A study comparing the lack of runs in Zener card sets with the lack of runs in guesses by humans also showed that they would produce enough apparent hits to match many of the Rhine trials.

We simply don't know how rigid the controls were in the Linzmayer series. Rhine remarks of the series where Linzmayer got 21 out of 25, "In this series he did not see the cards as they were being dealt and called." The implication seems to be that for the majority of the testing Linzmayer *could* see the cards. Unfortunately, Rhine was not able to go any further with Linzmayer, except for a separate hastily arranged series of 900 tests. As

Rhine puts it, "But (and alas!), after three most exciting days of experimentation with Linzmayer, he had to leave."

Rhine continued doggedly if somewhat randomly to engage in tests when he could get his hands on participants. The results were not up to Linzmayer's scores, but the majority did come up better than chance would suggest with a comfortable level of confidence. The trials never came up with the kind of result you would expect if you could actually see the cards, or read someone's thoughts—which in itself is interesting—where we would expect everyone to get many more right answers than wrong. But the subjects frequently did better than they ought to by chance alone.

Although Rhine seems to have set the trend of ignoring both emotional connection and stress as potential mechanisms for encouraging telepathy (see page 30), he did put considerable faith in the ability of financial encouragement to bring out the best in psi abilities. To put it bluntly, he believed that the offer of money focused the mind wonderfully, even though many might think that receiving cash for results would be a good incentive to cheat.

The most dramatic example of this effect was a test undertaken at the height of the Great Depression, when Rhine's subject that we met at the start of the chapter, Hubert Pearce, managed to name all 25 cards in a test run correctly. The stakes were high—Rhine had offered a massive $100 a card. Back then, $2,500 was an extremely generous amount of money, at a time when the average price of a house was around $3,400. Unfortunately for the financially struggling Pearce, Rhine claimed afterward that he had only been joking in offering such a large sum. Interestingly, Rhine did not mention these financial incentives when he wrote up the details of Pearce's work in his book.

The thumbnail sketch Rhine gives of Pearce is fairly straightforward. Pearce was at the time a young Methodist ministerial student in the Duke School of Religion. Rhine assessed him as

> very much devoted to his work, though fairly liberal in his theology. He is very sociable and approachable, and is much interested in people. There is also a pretty general artistic trend to his personality, expressing itself mainly in musical interest and production, but extending into other fields of art as well. . . . All of Pearce's work has been carefully witnessed; but I wish to state in addition that I have fullest confidence in his honesty, although in this work the question of honesty arises in my mind with every one, preacher or no.

It is be hoped that the question of honesty did arise and was kept under strict control, as once again the experimenters were using the early Zener cards with their high risk of being read from their backs. Even when efforts were made in the early Rhine tests to avoid the subject's being able to see the sender and the cards, there were some significant flaws. As we shall see, even apparently impressive results with sender and receiver in different buildings could easily have been obtained without any telepathy taking place.

Pearce's initial tests were unusually consistent. Across 2,250 trials he succeeded 869 times, meaning rather than achieving 5 correct guesses out of 25 he was getting around 9.7 right. With such a large number of trials this pushed the probability of guessing this number by chance up into the billions to one against. There were some oddities in the procedure. Rhine comments that in this series, Pearce did not seem to be helped if the sender looked at the cards. In fact, when the experimenter did so, Pearce's score got worse. It would seem that the cards were simply picked off a pile one at a time, but frustratingly, once again, we have no idea of the exact conditions that were involved. Pearce also proved able to force a low score—guessing wrong more frequently than chance predicted, something that Rhine found quite fascinating, often asking subjects to come up with both a "right" and a "wrong" answer, both of which deviated from expectation.

There is no doubt at all that if there is neither experimental error nor cheating in Pearce's work, then it does successfully demonstrate that psi abilities exist. He took part in a total of over 15,000 trials. In a range of comparable trials numbering 11,250 he achieved an average of 8.9 correct guesses out of 25, compared with a random chance probability of getting 5. The numbers of trials are so large that it makes the probability of this happening by chance ridiculously low.

Rhine uses an old-fashioned measure of chance occurrence called deviation over probable error. In effect he divides the distance between the actual score and the mean score by a value related to the typical width of the expected distribution. By his figures, the probability of this value being 8 is around 1 million to 1 (I make it 2.9 million to 1 with a modern method), and the value climbs steeply, so with a deviation over probable error of 9, Rhine reckons there is only a 100-million-to-1 possibility of this happening by random chance.

To emphasize just how remarkable Pearce's scores were, Rhine calculated that across those 11,250 trials his deviation over standard error was not 8 or 9, but 60—making the odds more than astronomical and certainly beyond the capabilities of my calculator. In practice, Rhine's statistics were a little flaky. He used a method that would work for random drawing from a pack of 25 with the card drawn then being returned to a shuffled pack every time, but because the actual experiment involved drawing a card, guessing, discarding it, drawing from the remaining 24, and so on, this makes the chances of something unexpected turning up better. But it doesn't stop the kind of scores Rhine was getting from being astronomically unlikely. These results could not be the result of chance.

What is without doubt is that many of Pearce's tests were carried out under poor control conditions. Let's revisit the test that opened this chapter. Pearce, looking slightly uncomfortable in a jacket and tie, sat at a small wooden table, directly opposite the experimenter, often Joseph Rhine himself or his assistant, Joseph Pratt. Pearce was allowed to handle the cards, and it was Pearce himself who took the cards off the pack one after the other as he guessed what the value was. Pearce also handled the cards when they were being checked against his predictions.

Of itself this doesn't prove that Pearce was cheating, but the conditions were such that it would be relatively easy for anyone with malicious intent to manipulate the experiment. Rhine claimed that "there is no legerdemain by which an alert observer can be repeatedly deceived at this simple task in his own laboratory." He brought in Wallace Lee, known as "Wallace the magician," to test this out. Although not exactly a Houdini, Lee was a popular school show magician who ran a mail-order business selling magic gimmicks. According to Rhine, Lee attempted to duplicate Pearce's work under the same conditions and was unable to score above chance, admitting that he didn't know how Pearce did it.

While this sounds relatively impressive, and we have no evidence that Pearce had any magic training, Wallace the magician's inability to duplicate the ability does not mean that Pearce was definitively not cheating. Wallace appears to have been primarily a school comedy act, and there is no evidence he specialized in close-up magic. I would be very surprised if James Randi could not duplicate Pearce's abilities given such freedom to handle the cards. I am no magician, but I can think of three or four ways to do this, the most obvious being to sneak a peek at an upcoming card when the observer was concentrating on recording the previous guess.

Pearce did not have to see every card. He had only to see around one in four to comfortably push his score up to the kind of level he was achieving. In this particular case, the person controlling the experiment did not look at the cards until after the guess had been made, but in other trials he or she was looking at the card to mentally send the value. As we have already seen, if this is done across a table, as was often the case in Rhine's experiments, and the sender wears glasses, it is not too difficult to consciously or unconsciously see a reflection of the easily distinguished shapes in the glasses. And if a would-be cheat can gain access to the cards before the test, it is all too easy to leave markings on the cards that are pretty well imperceptible unless you know what you are looking for. Wallace Lee might not have been able to duplicate Pearce's performance in a particular instance, but there were plenty of trials he could probably have worked to his advantage.

Not all of Rhine's tests were this sloppy, though a fair number were. In some tests more separation was put between the person tested and the cards, or a screen was used to block visibility, or cards and the subject were in different rooms to reduce the chances of accidentally discovering a value. However, this inconsistency is itself a concern in taking Rhine's results seriously. Why he was not more consistent in his controls is not at all clear, and this muddies the waters for anyone attempting to make sense of his data.

Pearce's parlor trick was being able to call out the cards in a pack in order before the cards had been dealt out, a technique that would not allow for the kind of cheating as you go that seems likely in the previously described experiments. He was, it appears, somehow scanning through the pack, definitely using remote viewing rather than telepathy because at this stage no one knew the order of the cards in the pack. These trials weren't quite as successful as his other attempts, but Pearce still managed to average 7.9 out of 25, as opposed to the expected 5.

Rhine makes one telling point. In this particular style of test, Pearce was better with cards toward the top and bottom of the pack and worse with the cards in the middle. It was these middle cards that dragged his score down from his typical score of about 10 hits per set of 25. The most obvious reason for this, which doesn't seem to have occurred to Rhine, was that Pearce had the opportunity at some point to take a quick look at the whole pack while not observed. If time was limited, it seems most likely he would glance at the first and last cards, which would be easier to remember in sequence than the center cards of the pack.

In another series between 1933 and 1934, Hubert Pearce was put through remote tests, with impressively arranged controls against cheating, that continued to produce results that were far better than could be expected from random chance. This set of tests, often cited as the Pearce-Pratt experiment, is usually considered one of the strongest pieces of evidence ever recorded under controlled conditions for the existence of clairvoyance or remote viewing. This is very much the gold standard from the Rhine lab.

To avoid any chance for the receiver to observe the cards, the sender, Joseph Pratt, remained in Pratt's office at the top of the physics building (where the Psychology Department had a small outpost). He and Pearce synchronized their watches in a move reminiscent of early war movies; then Pearce crossed to the library building and sat in a cubicle in the stacks. Pratt was able to observe Pearce crossing the quadrangle to the library from his office window. Now Pearce was over 100 yards away, on the opposite side of the large library building with the wide-open grassy space of the quadrangle between them.

Pearce achieved some remarkable results. Each minute Pratt would take a new card from a shuffled pack and place it facedown on a book. He did not, apparently, look at the face of the card at this point, but simply selected a new card every minute. Working purely on timing, the pair then ran through a set of cards. After each sixty seconds elapsed Pratt would remove the card from the book and replace it with the next from the pack. When the first pack of 25 cards had been used there would be a five-minute break, after which a second pack was worked through. At the end of the series Pratt went through all the cards and noted down their values, and then each participant provided a sealed record of his observations to Rhine.

The distance was then increased to 250 yards, with Pratt moving to the university medical building, a couple of blocks farther away from the library. It apparently took Pearce some time to adjust to the extra distance with an initial period of low scoring (as also occurred when they first tried the 100-yard separation), but over time he began to pick back up to his typical average of just under 10 correct guesses out of 25. Rhine believed that this showed that such mental feats were not influenced by distance, though this kind of range is not great enough to deduce this definitively.

Overall, at both distances, Pratt and Pearce undertook a total of 37 trials—1,850 guesses, which was plenty to provide good evidence and to make scores of just under 10 out of 25 pretty well impossible to envisage based on random chance alone. It would be expected to occur randomly around once in every 10,000,000,000,000,000,000,000 attempts. You are more likely to win a major lottery every day for your entire life.

It sounds like an impressive test, but according to Professor C. E. M. Hansel of Manchester University, who performed a detailed analysis in the 1960s of a range of classic psi experiments, there was a huge flaw in the way that the test was actually carried out. As always, the devil was in the detail. There seems to have been no attempt made to ensure that Pearce really was where he was supposed to be. The most basic of precautions—having someone else keeping an eye on Pearce during the experiments—was overlooked.

When Hansel visited Duke University as part of his research he found it easy to take a sneaky look into one of the department's offices without a researcher's knowledge. In fact, to demonstrate just how possible this was, he did a trial himself under similar conditions to the Pearce long-range test with W. Saleh, whom Hansel identifies only as "a member of the research staff at Duke." By peeking over the crack at the top of the door, Hansel was able to get 22 correct values out of 25 as Saleh made a record of the cards used after the guessing phase of the experiment. In Hansel's discussions with Pratt it also turned out that Pratt did not lock his office door and may have left the cards on his desk in the same order as they were in during the trial after he left the office.

In the original experiments there was no attempt to shield Pratt from visibility while he was going through the cards to record the values, nor was there any attempt to ensure that his office was inaccessible. While neither this potential access nor the lack of supervision of Pearce means for certain that there was any cheating, the important fact that Hansel uncovered was that there was every opportunity for cheating to take place—and in these kinds of experiments, opportunity is enough to make the results worthless.

What could have been a very valuable experiment was ruined for the want of a few simple precautions. It would have been easy to supervise Pearce, and to check the cards in a way that would not have been visible from outside the room. Even if Pratt had turned around at the table so that he faced the door it would have helped immensely, as it would have made it much harder to tell what the cards were from the corridor. Combine this with shielding the table from view of the windows and door, as well as locking the room afterward and shuffling the cards immediately after the values had been noted, and the experimental setup would have been practically foolproof. But none of these things seems to have occurred to Rhine or Pratt.

It might seem unlikely that Pearce could successfully cheat on so many test runs—there were 74 sets of 25 cards in all. But in practice he didn't need to cheat every time. When each of the results of the runs is looked at independently, there were a total of 20 tests where he scored between 4 and 6 out of 25—the most likely scores if Pearce had simply guessed. If you permit a little more flexibility and allow him to have randomly guessed between 3 and 7, then 29 of the 74 sets reflected pure guesswork.

Pearce did not have to cheat successfully every time. And as Hansel points out in his analysis of the experiment, because it ran on a strict timetable, Pearce knew exactly what Pratt would be doing when, and knew when to peek into Pratt's office if he wanted to watch as Pratt noted down the card values.

As I make clear in chapter 10, there is good evidence that some subjects will cheat, given the opportunity, so it is essential that all chances of gathering information through the conventional senses be removed. There is also some concern with this particular series—which, remember, is generally considered among the best scientific evidence there is for clairvoyance—considering the accuracy of the recorded data. Pratt and Rhine wrote up the experiments for many different publications over the years. There are at least five variants of what the actual scores were, listed run by run in these different publications. Five copies of what is supposed to be the same data, each listing different values. While this does not necessarily suggest cheating by Rhine and Pratt, it does demonstrate a rampant carelessness with the data that is hardly encouraging if we are to take these experiments seriously.

We don't know how all of Rhine's distance tests were carried out, though he states he tried varying distances, moving the receiver twelve feet and then twenty-five feet away, with walls in between sender and receiver. This was done to eliminate a particular concern, "unconscious whispering," an unsubstantiated theory popular at the time positing that people would say subvocally what they were trying to communicate mentally, and that this might carry at the audible level. A telegraph key was used for signaling in some cases, with the sender using this to alert the receiver that he or she was about to send an image. Rhine was of the opinion that these tests demonstrated that some aspects of psi ability got stronger with distance, "contrary to what wave mechanics would lead us to expect," but in practice the results he recorded aren't sufficient to make this assertion.

Rhine did also give Pearce some "pure telepathy" tests in which the sender simply chose a shape at random from the five available and projected it. In this series of trials Pearce began pretty much at chance levels of success, but over time his scores edged up to around 7.1 out of each 25—nowhere near his success with the actual Zener cards, but still respectably above the random expectation of 5.

This sounds very impressive. Hansel's detective work is not relevant here, as there was nothing for Pearce to see, so he couldn't sneak up and peek. But at the risk of sounding like a broken record, without more details of the controls imposed it is very difficult to be sure that the test was proof against cheating or accidental distortion of values. There are a number of danger points I can envisage, though I am sure a James Randi could come up with more.

First and foremost, there could be bias in the "random" selection. The sender couldn't use a random table of Zener images, or random numbers between 1 and 5, because the point where he looked at this table would be an opportunity for remote viewing, and Rhine hoped to eliminate this as a possibility. So the sender had to pick symbols "randomly" off the top of his head. Unfortunately, as we have seen, human beings are hopeless at generating a stream of random numbers or images. We really can't envisage randomness and don't repeat the same value nearly often enough in a sequence. If the receiver merely assumed correctly that the next image was different from the previous one, he immediately went from getting 1 in 5 right to 1 in 4 by pure guesswork.

Rhine pointed out that the sender had "freedom to repeat or vary in any conceivable way," but the mere existence of that freedom is worthless because human beings are very bad at using it. The fact is that people rarely come up with the same item twice, and it is even more unusual for them to dream up a run of more than two in a row. If someone was trying to imagine a random stream of heads and tails, it is very unlikely she would come up with the H T H H H T H H T T sequence my coin produced. If a sender typically went through all five Zener symbols in a telepathy test, and the receiver was being told whether he achieved a hit or a miss, then by using a simple rule of "Stick with your guess until you are correct, then change," once there had been a success, the chance of getting another hit would get higher and higher.

The closest we can achieve to randomness in a trial where the sender is not using any aid to come up with his "random" choice is for the sender to think of a series of random words. These won't truly be random either, but the chances of guessing more than one or two by pure chance is so small that such a performance would be highly telling—and with thousands of words to choose from, the chances of a repetition being expected in a random sample become much lower.

Something we don't know about the pure telepathy trial is how the sender recorded the values that he or she was sending. It would be pretty well impossible to accurately remember a sequence of 25 values in a run. How did the sender do this recording? As soon as something was written down, it becomes possible to use remote viewing (or, mundanely, to find a way to peek).

We don't know, for that matter, how Pearce recorded his guesses. Did he write them out himself as he made them, or did he call them out and have someone else note them? Was there any chance to tweak these values toward the actual sequence of symbols after the event when it was possible to find out what had been sent? It certainly wouldn't be unheard of for subjects or experimenters to make changes to the data to better match their expectations.

Finally, we always have to ask if there was any selection. As we have already seen, as soon as you have the opportunity to decide which trials you will count and which you will ignore, there is a chance to shape the results. Either the experimenter or the subject can ask for this. The subject can say after a bad run, "Can we not count that? I was disturbed when someone came into the room." Or the experimenter can decide for some reason that the run was anomalous and shouldn't be counted.

I'm sure that's not comprehensive, but it's a start. Apart from the Pearce experiments that Hansel demonstrated could have been cheated if Pearce did not stay in the library, Rhine generally found that telepathy worked reasonably well at a distance, but clairvoyance didn't. He attributed this to there being different mechanisms for the transmission of information, with clairvoyance being better at short range—but it seems more likely that the successes in clairvoyance were the result of being able to see the cards, while the successes in telepathy originated more from the bad experimental design that required the sender to dream up a random sequence of shapes, to hear the results (and not be influenced by them), and to remember the outcomes in batches of five before recording them.

To make things more complicated, Rhine also introduced a number of variables into the procedure to see if they had any effect. As well as running an electric fan at times to cover that involuntary whispering that was assumed could ruin things, he submitted some subjects to hypnosis to see if this had any effect on their ability. (It didn't.) He also tried dosing subjects with sodium amytal to make them drowsy (usually reducing scores) and caffeine to make them more alert (usually giving a small but significant boost).

There is something magnificently amateurish about these attempts to try out the impact of random doses of stimulants or sedatives. It bring to mind the experiment in August 1962 when two doctors from the Oklahoma School of Medicine, Louis Jolyon West and Chester M. Pierce, decided to try out the effects of LSD on an elephant. They obtained an elephant by the name of Tusko from the Oklahoma City Zoo and, having no idea how much to prescribe, gave him a vast amount of the psychedelic drug, pumping in around three thousand times the typical human dose. The hope was to see if there was an induction of psychotic behavior, which was a concern for those using LSD—but the actual outcome was that Tusko collapsed, his tongue turned blue, and within two hours he was dead.

It's not that Rhine was putting his subjects at such a huge risk as Tusko suffered with his chemical tinkering, though the drug sodium amytal, also known as amobarbital, at the time considered to be a "truth serum" (this has since been disproved) as well as a sedative, was not without its dangers if improperly administered or combined with alcohol. But the way he played with these additional features in his tests suggests that same spirit of whimsical discovery that was all too obvious in the "elephant on acid" experiment.

Rhine made thousands more trials often with seemingly impressive results, but always with the specter of doubtful experimental protocols hanging over him. In his reports there are frequently tantalizing suggestions that something would benefit from further investigation to check just how a result was obtained. And there is no better example of this than the work of a Mr. George Zirkle, something Rhine comments on very favorably.

Rhine tells us that Zirkle, a graduate assistant in his department, achieved remarkable results in communicating telepathically with his fiancée, Miss Sarah Ownbey, one of the department's graduate students. One notable point was an excellent success level, averaging 11 hits out of 25 over 3,400 trials, and featuring some particularly spectacular runs. "Zirkle's scoring," comments Rhine, "is sometimes phenomenal. He several times has gotten 22 in 25 correct. Once in calling 50 in a series, 26 were found to be correct in unbroken succession." The other significant point in these experiments is that Zirkle proved hopeless at clairvoyance tests, where there was no sender.

One possible interpretation of these results, as was observed in chapter 3, is that here at last were two subjects who had the kind of bond that is often associated anecdotally with telepathy. It seems very strange that researchers usually don't give more consideration to the possibility that two individuals who have a strong connection, whether biologically or personally, may produce better results. Alternatively there is the dark side of interpreting results from a pair who knew each other well.

If there was cheating going on that involved both parties, it was clearly more likely to happen if the subjects were well acquainted, something we can assume from the engagement (Rhine tells us that Sarah Ownbey did eventually become Mrs. Zirkle). And even if Ownbey was not involved directly, the chances are that Zirkle would have a better chance to gain peeks at the cards when working with his fiancée. One obvious possible reason for the "peculiarity" Rhine raises that Zirkle was great at telepathy but was unable to work clairvoyantly is that it would be harder (though by no means impossible) to cheat with a remote-viewing experiment, where no one knew what the cards were in advance, than with a telepathy experiment, where they were faceup during the process.

It is worth briefly mentioning one other test series undertaken by Pratt, as it is often cited alongside the Pratt-Pearce experiment as a near-perfect example of confirmation of clairvoyance. In this Pratt-Woodruff series, undertaken in 1938 and 1939, significantly more effort was put into controlling the subject. Although the active person and the operative with the cards sat at the same table, they had a screen between them that prevented the person doing the guessing from seeing the (facedown) cards. To try to avoid experimenter bias, the experimenter could not see which result the subject selected. When the subject made a selection she pointed to one of five blank cards, which lay beneath a set of five key cards only she could see. This meant the person recording the results was blind as to whether the guess was right or wrong until later in the proceedings, when the value had already been written down.

This series came up with positive results across 60,000 trials. Although the outcome was only slightly different from the expectation of chance—averaging 5.2 out of 25 where 5 would be expected, which required just 1 extra correct guess out of 100 trials—because there were so many trials it was a million-to-one chance that this outcome would occur. And the test was painfully close to being well controlled. But there was still at least one big problem.

If only the person guessing and the experimenter or "operator" with the cards had been kept in separate rooms without any way of communicating during any particular set of trials, and if the results had been compiled separately, there would have been practically no opportunity for cheating, whether intentionally or unconsciously. But putting the two individuals on opposite sides of the table made it possible for the person operating the cards to gain some idea of what the other person was guessing and so to remove the blind aspect of the trials, meaning that the person handling the cards could bias the outcome.

The way the setup was supposed to prevent this was that after each run, the guesser was supposed to randomly rearrange the key cards corresponding to the blank ones visible to the operator, after the operator had seen the current sequence to note the results. This rearrangement was done behind the screen. But there were two problems. One was that the key card order was frequently not changed from run to run; the other was that even when it was changed, the key cards don't appear to have been shuffled, and the way that the cards were rearranged, it should have been relatively easy to guess what the first and last card in the key sequence was. When the data was subsequently analyzed, the guesses corresponding to the first and last cards were much less likely to have occurred randomly than the rest, suggesting there was some information leakage for these cards.

Once again, what could have been a real opportunity to show make-or-break results was let down by flaws in the experimental design.

There is no evidence that Rhine himself was involved in anything fraudulent, but the same cannot be said for at least one of his associates. In 1974, the then director of the Rhine Institute, Dr. Walter J. Levy, was forced to leave the institute under a cloud.

Levy had been working on a series of experiments that seem, frankly, bizarre at first sight. With all the possibilities for parapsychology research on humans, Levy chose to observe the psi capabilities of chicken eggs. It's possible to think of justifications for the ability of a human brain to produce something like telepathy, but there was no brain present here. This wasn't even a chick with a fully functioning nervous system—the eggs were fertilized but were still at the incubation stage. This state, in fact, was crucial to the reasoning behind their assumed psi abilities.

Eggs in an incubator benefit from the warmth of a heat lamp. In Levy's test, the lamp was turned on and off by a random time switch. In principle, it should have been off around half the time, but in practice it was found to be on more frequently than it was off, suggesting (according to Levy) that the eggs were influencing the experiment to make conditions better for their development. When the same experiment was tried with hard-boiled eggs (this experiment is very difficult to report with a straight face), there was no influence on the time switch. If ever there was an experiment demonstrating that at least some psi effects are the result of statistical anomalies or cheating, it has to be this one.

Levy went on to perform a similar experiment with animals that did at least have a functioning brain, though the ethics of this experiment would raise eyebrows these days. He wired up rats to a device that randomly fed currents to their brains to generate a pleasure response. Once more, the positive outcome occurred more often than simple chance would allow, as if the rats were influencing the device to give themselves more jolts of gratification, though this still occurred in only 55 percent of trials.

The good thing about not using human subjects in these trials is that it took away the potential for deception. While rats are perfectly capable of cheating given the opportunity, this was a circumstance where there was no chance for them to do anything but act as required. While it is arguable that, if psi abilities exist, there may be a need for a human brain (or at least *some* brain) to make it feasible, at least these trials should have been free from the suspicion of cheating by the subjects. No magicians were needed here to see if the subjects were fooling the scientists. Unfortunately, though, there was no provision to check up on the scientists themselves.

Levy's colleagues, perhaps a little surprised by the psi abilities of eggs that had yet to form a nervous system, kept a hidden watch on his experiments and witnessed the scientist pulling the plug out of the recorder on occasion to prevent it from making a mark for a negative result. When a second recorder was secretly connected, while Levy's recorded results continued to come out well above random, the secret recorder showed that there were no significant incidents. Confronted with the evidence, Levy confessed that he had felt pressure to come up with a positive outcome—the downside, perhaps, of following in the footsteps of someone with Rhine's reputation—and had falsified the results to make sure that he did not fail.

Could such deception by the experimenters have happened with others in Rhine's lab as well? Was there indeed the kind of fraud that has been suggested on the part of top performers like Hubert Pearce? It is impossible to be definitive, but there do seem too many doubts about the experimental controls that Rhine used to have a huge degree of confidence in their accuracy. Taken at face value, had proper controls been imposed, the results were astounding. In their initial three-year period, the Rhine lab staff made an astounding total of 91,174 trials. With such a large number of tests, even a relatively small deviation from the expectation of random guessing would have made the outcome highly significant.

As we saw earlier, Rhine's measure of choice for the probability of something happening by chance was called deviation over probable error. If this measure had a value of 8, he reckoned the chances of this happening without any "steer" from psi ability or cheating was around 1 million to 1, while with a value for deviation over probable error of 9, Rhine reckoned there was only a 100-million-to-1 possibility of this happening by random chance. Over 91,174 trials, he calculated that the observed results would only have occurred randomly with a deviation over probable error of 111.2. You would wait for longer than the lifetime of the universe for that to happen spontaneously. It was about as likely as the also theoretically possible but highly unlikely chance that all the atoms in a car would jump sideways out of a garage simultaneously, leaving the car sitting in the driveway.

Did Rhine himself act fraudulently? While this is possible, it seems doubtful. He himself discounted this with the stuffiness of the period by saying "an academic person has seldom, if ever, been found to work a deliberate hoax involving hard work and long hours for several years." Unfortunately, he was wrong. Academics have, quite a few times. Since Rhine wrote that, there have been famous examples of exactly such a "hoax"—and Rhine was not, of course, aware of Levy's work, which would come much later. All the evidence is, though, that Rhine himself was honestly attempting to do serious research, and as he pointed out, the work was collaborative enough that one person's attempts to fake things were likely to get noticed by his or her peers.

What is much more likely is that a number of factors conspired to produce Rhine's remarkable results. Some of them were, in all probability, the result of mere chance. Some of those taking part as subjects—Hubert Pearce, and Sarah Ownbey and George Zirkle, spring to mind as obvious examples—could well have been cheating at least some of the time. Similarly, some but not all of Rhine's assistants might have felt the urge to tweak results to get something positive out of their long effort (and perhaps even to ensure they made the grade for their degree or doctorate). Throw in the sometimes loose approach to statistics and the obvious inclination to cherry-pick as much as possible that comes through occasionally in a concealed fashion in Rhine's documentation of his work, and it is, unfortunately, possible to cast doubt on the overall impact of Rhine's work.

Taken in isolation, despite these concerns, it is easy to be swayed by the sheer volume of Rhine's tests. Surely with so many experiments, something was proved? But you can continue using a flawed test as long as you like and you will continue getting flawed results. In ordinary science, the real proof is not when one laboratory has produced a vast series of results, but when others manage to duplicate those results. What is frustrating is that it is quite possible that Rhine did witness genuine telepathy—we just can't be sure from the results he published.

After Rhine produced his book on the experiments there was a huge amount of public interest, and a number of other universities duplicated his experiments, but with stricter controls in place. These weren't casual affairs, or quick and dirty attempts to reproduce some aspects, but long trials on a comparable scale to Rhine's. At Princeton 25,064 trials were run; at Colgate University, 30,000 trials; at Southern Methodist University, 75,600; and at Brown University, 41,250. Perhaps most impressive was a series of 127,500 tests at Johns Hopkins. These combine to provide a serious reference point for Rhine's work. And not one of them produced comparable significant results.

The Rhine series of tests were interesting. They were a noble effort. And they have become so iconic that over eighty years later, they are still regarded as one of the high points of the investigation of ESP. But the relatively amateurish way the program of experiments was run ensures that Rhine's work cannot give us the definitive proof of the existence of telepathy and clairvoyance that he hoped for, and the subsequent failures to reproduce his results in experiments that were intended to support him, not to undermine him, suggests that it is not sensible to give any great weight to Rhine's work.

Nonetheless, Rhine's work did certainly have one important outcome. When the U.S. military became suspicious that the Soviets were busily at work employing ESP agents to spy on and interfere with the West, they took the threat seriously. With all that data from Rhine's work at Duke, the powers that be were not going to ignore the possibilities of psi wars.

8.
ENTER THE MILITARY

Imagine the scene. A military office—a large office. This is not the cubbyhole of a lowly quarter sergeant; it is the office of a general. The office's occupant stands with his back to the wall, facing across perhaps twenty feet of space toward the opposite side of the room. He is preparing himself mentally, visualizing the wall in front of him and its atoms. He knows that most of that wall is empty space. It's not just that there are gaps between the atoms—every single particle of matter is mostly empty space with almost all its mass concentrated into a tiny nucleus. Compared to the size of the atom as a whole, the nucleus is like a fly in a cathedral. It should be easy enough, with focused willpower, to brush past a collection of flies.

The man tries to clear his mind, to picture the atoms in his own body slipping through those of the wall. And then he begins to move. What starts as a walk develops in a pace or two into a run, straight toward the opposite wall. He has total faith in his ability to pass straight through the barrier. Or at least he thinks he has. With a crash and a stinging pain in his nose he is brought up short. Once again, the wall is not going to let him through.

This was not some student high on psychedelic drugs. This was Major General Albert Stubblebine, at the time in charge of the entire Military Intelligence Corps. Stubblebine had become convinced that the human mind was capable of far more than many scientists would give it credit for—that his own mind should be able to influence matter. Inspired by a Defense Intelligence Agency report on Soviet activity in the field of psi warfare, Stubblebine was not going to let the United States be left behind.

Inevitably, it wasn't just academics who became interested in psi phenomena in the twentieth century. When that visionary friar Roger Bacon suggested back in the thirteenth century that it would be valuable to see remotely into a citadel, he had military applications in mind. He knew that being able to take a remote view of the enemy was an essential component of a successful military campaign. Similarly, when Galileo first demonstrated his telescope to the Doge's Council in Venice, there was no doubt it was the defense of the city that gripped the imagination of those dignitaries, not the device's potential for stargazing. Remote viewing is a natural tool for military surveillance, and with other psi abilities, in the 1970s it came under scrutiny from the U.S. Army.

This was a topic that was in the air with the media coverage given to the Geller phenomenon (see chapter 10), but there was also an element of Cold War one-upmanship. The United States was still smarting from the early lead that the Soviet Union had taken in the space race. In the end, superior U.S. technology and skill had won through, enabling Americans to reach the moon first, but before then there had been the embarrassing sight of Sputnik declaring Soviet leadership in space, followed by the Russians getting the first man into orbit. The powers that be were not going to leave the Soviets to dominate any aspect of life with the potential to improve military superiority.

In the early 1970s, just before Uri Geller reached the big time, the Defense Intelligence Agency had produced a report titled *Controlled Offensive Behavior—USSR*. It sounds like a study of missiles and troops, poised to invade the West, but this intelligence report tried to pull together any information that had been gathered about Soviet attempts to use psi abilities for military purposes. This fascinating document has now been declassified.

The report is primarily concerned with Soviet research on altering human behavior, so you would expect to find a study of the use of brainwashing, drugs, isolation, and other means to achieve behavior modification—and these are all present in the 173 pages of typewritten script. But what is particularly relevant to the study of psi abilities is that the report goes beyond the physical to what it refers to as the use of psychophysics. In the report's summary, its author, Captain John D. LaMothe, comments:

> The Soviet Union is well aware of the benefits and applications of parapsychology research. . . . Many scientists, U.S. and Soviet, feel that parapsychology can be harnessed to create conditions where one can alter or manipulate the minds of others. The major impetus behind the Soviet drive to harness the possible capabilities of telepathic communication, telekinetics and bionics is said to come from the Soviet military and the KGB. Today, it is reported that the USSR has twenty or more centers for the study of parapsychology phenomena, with an annual budget estimated at 21 million dollars.

The Soviets, opined Captain LaMothe, had been working on this clandestine mental technology since the 1920s and with this head start and superior financial backing were likely to have far greater knowledge than any programs that had so far been initiated in the United States. The report strongly implied that psi abilities were simply a fact and were being used as everyday tools of intelligence gathering in the Soviet Union. "The Soviet Union," commented Captain LaMothe, "is well aware of the benefits and applications of parapsychology research." There is no suggestion here that there is any doubt about the existence of any of the psi abilities.

Ironically, it seems that the Soviet investigation into the paranormal was kicked into overdrive by a fictional account of U.S. telepathic activity. In 1960, French journalists publicized a rumor that the crew of the U.S. submarine *Nautilus* were using telepathy to keep in contact with the shore. "Is telepathy a new secret weapon?" the French articles asked. "Has the American military learned the secret of mind power?" Although there was no actual basis for these stories, they were enough to trigger action by Soviet scientists. Combining this apparent U.S. success in using telepathy for military purposes with anecdotal evidence of earlier Soviet trials in telepathy—in all probability far worse in their controls than those of Joseph Rhine—was enough to start the ball rolling.

According to the report, a Soviet researcher, L. L. Vasilev, "claimed to have conducted successful long-distance telepathic experiments between Leningrad and Sevastopol, a distance of 1200 miles, with the aid of an ultra-short-wave (UHF) radio transmitter." It's not clear what LaMothe is suggesting here. How could this be telepathy while using a radio? (Perhaps the radio was merely used to set up the experiment.) It also seems that the Soviets were considering telepathy to be a suitable backup for radio when communicating with their space capsules.

The report then goes on to discuss an aspect of psi that is not covered in this book, because it is one of the supposed abilities that seems far-fetched enough to be well beyond the bounds of likelihood, and that tends to feature more in the claims of stage performers than in academic parapsychology. This is the ability to "apport": to move a physical object from one place to another without its passing through the space in between—in effect, a mental version of a Star Trek transporter. Although Uri Geller has claimed to have this ability, it is not one that he has ever demonstrated under even the weak laboratory controls that he has typically allowed, and even Captain LaMothe seems a little uncomfortable with the concept.

"The following discussion on apports and astral projection," he comments, "is not intended to be an endorsement for its scientific verification or even its existence. However, reputable scientists in the USSR and the U.S. are keenly interested in this phenomenon." The report then goes on to list plenty of anecdotal examples of these alleged phenomena without any criticism or comments on the degree to which fraud or simple storytelling could be involved in what seems little more than fantasy. It is worrying indeed when a military scientific report is based on such unquestioned material. It doesn't really matter that there was the disclaimer at the start—the mere fact that LaMothe detailed these goings-on would be enough to give them some credibility and to encourage funding of research.

It is enough to note that the prime authority cited in the report is the scientist William Crookes, who frequented séances and seemed to have little idea of the possibilities for deception.

What Crookes described was the typical work of the fraudulent medium around the end of the nineteenth century and into the early twentieth century, the kind of act that had been debunked time and again by the time this report was written. Again La-Mothe's discomfort shows through when he writes, "If any of this highly questionable material is true . . . ," but his attitude is that the subconscious mind is very powerful (the basis for this statement is not clear) and that parapsychological phenomena are caused by the subconscious mind, so such bizarre occurrences may be possible and are worth investigating.

LaMothe goes on to describe experimental work with a Soviet woman named Nina Kulagina, who "reportedly moves objects by sheer will." Mrs. Kulagina supposedly displaced items mentally and separated the yolk from the white of an egg placed in a sealed aquarium six feet away from her; but her pièce de résistance seems to have been that she managed to stop the beating of a frog's heart in solution and then reactivated it. As LaMothe drily remarks, "This is perhaps the most significant PK (telekinesis) test done and its military implications in controlled offensive behavior, if true, are extremely important."

What remains massively in doubt is if any attempt was made to prevent cheating in any of the examples in this report. As we have seen with the Rhine experience, even in a U.S. university, experimental controls could be extremely lax. What really happened in the Soviet laboratories will never be known for sure, and certainly cannot be considered reliable evidence. In another DIA report from the same period on parapsychology research in the Warsaw Pact countries, the redacted authors comment, "Most of these data, however, are difficult to adequately evaluate. Usually, sufficient information on the experimental procedure is not clear, or the number of experiments reported is very small."

All the evidence is that the work in the Soviet Union that inspired the U.S. military's interest in parapsychology was on a level with the psychic phenomena that had so intrigued nineteenth-century scientists: badly controlled, very limited in data, and with negligible credibility. There is a very strange pair of statements from LaMothe about observations made of telekinesis star Mrs. Kulagina by Western scientists, including the Rhine lab's Joseph Pratt. He comments: "Observations by Western scientists of Mrs. Kulagina's PK ability has [*sic*] been reported with verification of her authentic ability. These same Western scientists have reported that, as of February 1971, they have not been able to visit or observe Mrs. Kulagina."

At first sight this is contradictory—the scientists have somehow verified Mrs. Kulagina's ability without visting her or observing her, though presumably what it means is that they were given some access, but then it was withdrawn. LaMothe goes on to wonder why such a veil of secrecy has been placed around Mrs. Kulagina. It seems likely that even a relatively gullible observer would be able to discover how Mrs. Kulagina was obtaining her results, and that this did not involve telekinesis.

It's ironic that the the U.S. military was prompted to act in response to Soviet work that was itself inspired by a fictional account of American military use of telepathy. Whether or not there was any valid reason for responding, the CIA and others in the United States were unable to resist the feeling that they were being left behind. In 1972, the Stanford Research Institute lab that gave huge credibility to Uri Geller (see page 228) got a visit from a CIA operative. Aware of the apparent success in the

Soviet Union, the CIA felt it was important that the United States have its own research into psi abilities and provided significant funding, with an initial payment of $50,000 to support this work. The main outcome seems to have been the remote-viewing experiments described on page 125.

Neither these nor Major General Stubblebine's ideas, which also included the possibilities of psychic healing and of stopping the heart of an enemy by mental power, were the only work going on in the military. In fact, in the 1970s and 1980s, again perhaps inspired by the DIA reports, there were several attempts to use the apparent potential of psi for warfare. Unfortunately, there was very little selectivity among those looking for a parapsychological weapon. A typical approach would to be to take in every possible New Age idea on offer and give it a try. You have to admire the tenacity of the various army units involved, but some of these concepts were way off the scale in terms of quackery and pseudoscience.

Uri Geller has always claimed that he was hired by the U.S. government as a psychic spy in the 1970s. His claims were largely dismissed by the media, but given some of the more extreme areas of New Age thinking and wacky pseudo-technology that were tried out, it would not be at all surprising if this was true. This is the same environment that produced the episodes that were later fictionalized in the movie *The Men Who Stare at Goats*. A martial arts trainer was brought in to provide a mix of mystical guidance and straightforward unarmed combat training with the aim of producing a group of supersoldiers.

When the trainer Guy Savelli was interviewed by Jon Ronson for the factual book on which the movie was based, Savelli claimed to have both disabled and killed a goat with his mind—but this was nothing more than an anecdote, presented in a way that makes Savelli seem anything but a reliable witness. And this seems to be the problem with practically all the military attempts to harness psi, except where they were simply involved in funding university work. They produced vivid anecdotal results, but nothing concrete.

A good example of these strange, unsubstantiated stories is the reports of military remote viewing, an exercise sometimes referred to as Project Stargate. A small number of individuals, including Ingo Swann (see page 123), were employed for years to attempt remote-viewing surveillance. Some handled fairly mundane if slightly eyebrow-raising work, like attempting to monitor the whereabouts and actions of Panamanian dictator and CIA bogeyman General Manuel Noriega, or simply attempting to report on what was happening in territory the United States was interested in for military reasons. Others produced intelligence that was far more dramatic.

There seems to have been little to confirm that what was described in these remote-viewing sessions was anything other than fantasy. This became particularly obvious when the work of one individual, Ed Dames, became public. Dames was briefly infamous after leaving the military in the 1990s. He appeared in the media telling of dire threats that he had foreseen while remote viewing, such as three-hundred-mile-per-hour winds that would ravage America, and a canister containing a plant pathogen than he believed an alien race had attached to the approaching comet Hale-Bopp. Sadly, this last prediction, combined with a faked photograph that appeared to show a vast artifact bigger than the Earth tracking alongside the comet, seems to have been the trigger for the Heaven's Gate cult's mass suicide.

When I started researching this chapter I expected it to be one of the most important ones in the book. Surely the amazing potential for the military applications of psi would make the studies undertaken rigorous and definitive. Yet the whole military parapsychology investigation was conducted with an amateurishness that makes the controls used in a Uri Geller stage show seem effective. The military's attempts to take on psi seem to have lacked any pretense of scientific accuracy and objectivity and to have been devoid of precise laboratory controls. It was very much a case of "Try everything, give it a go, and either it works according to the way we feel it works, or it doesn't." And all the evidence that has been published is purely anecdotal. A conspiracy theorist could have a field day, suggesting that the reason the evidence is so bad and the anecdotes are so silly is that this is the result of a massive cover-up. But in reality what comes through at every turn is incompetence. The military work was just as lacking in value for a serious understanding of whether or not psi abilities exist as the accounts of the feats of nineteenth-century spirit mediums.

This was quite the reverse of the work we will meet in the next chapter, where a major university was persuaded to take ESP to its scientific limits. Think what you may of the results, there was no lack of objectivity and scientific control in the Princeton PEAR project.

9.

PICKING THE PEAR

A young woman, a student at Princeton University in New Jersey, sits on a comfortable seat. She might be in a students' common room, judging from the institutional row of plush sofa seating, the overly arty misshapen coffee table in front of her, and the uninspiring prints on the pine-paneled wall. Yet there are other indicators that maybe this is her bedroom. There's a stack of books in a small bookcase to her right and a whole pile of cuddly toys ranged along the top of the seating and stacked in a haphazard pile in the corner. To confuse the picture, right in front of her is a device that seems more like a feature of a bingo caller's nightmare than anything you would find in a campus residence.

A glass-fronted, wall-like machine stretches from floor to ceiling. In the top of the machine sit rank after rank of table tennis balls. In the lower half, arrangements of plastic sheeting allow the balls to fall down different paths and to accumulate in bins at the bottom. Time after time, a ball falls, bounces from sheet to sheet, and comes to rest. Over time a distribution grows in the positions of fallen balls, a real-life demonstration of a bell curve, with most balls in the middle of the machine. The woman concentrates furiously. Can she change the outcome of chance? Can she use telekinesis to force more balls to go to the left than the statistics predict? It's a battle of mind versus machine.

If Rhine's work in the 1930s marked the move from psychic research in the wild to psi studies in the lab, the other particularly outstanding academic effort to explore psi phenomena would take place at Princeton University in the form of PEAR—the Princeton Engineering Anomalies Research Laboratory. Just as physicists researching time travel tended to write about "closed time-like loops" in the early days to reduce the potential for academic scorn to be heaped on them, PEAR's title deliberately concealed its interests behind the bland topic of "engineering anomalies."

Most of PEAR's experiments were like the one described in the opening of the chapter, though many used electronic devices rather than anything as crudely constructed as the "random mechanical cascade," as the ball machine was called. The aim was to try to influence these devices away from their expected random behavior in directions that were pre-established before the experiment began. (Clearly it wasn't good enough to say after the event what your intention was, as it would be easy enough to match this to what had actually happened.)

For twenty-eight years from its founding in 1979 by Robert Jahn, the dean of the School of Engineering and Applied Science, PEAR's team, comprising psychologists and physical scientists, undertook thousands of experiments involving millions of trials to build a database of outcomes that could be analyzed using whatever statistical techniques could be brought to bear on it. Unlike the ganzfeld experiments (see page 67) this was not a program that could be accused of having unfortunately small sample sizes. At PEAR, quantity of data was crucial.

The outcome was what appeared to be a very significant confirmation of a particular kind of telekinesis. We've got to be a little careful here, because that high significance is the result of getting lots of *very* small deviations from the expectations of random chance—just a few parts in ten thousand; but as we have seen before, having so much data makes even small values very significant. While the smallness of the variation from chance, and the nature of some of the effects studied, raise some concerns, there is no doubt that PEAR did produce some results that were worthy of further study. This is no collection of unsubstantiated anecdotes.

The PEAR program also involved some remote-viewing experiments similar to, though hopefully better controlled than, those used by Puthoff and Targ at SRI (see page 116). This part of the program was a much smaller-scale undertaking—unfortunately these kinds of trials are much more time consuming—but PEAR accumulated around 650 results, again suggesting to those running the trials that something was happening that went well beyond the expectations of chance (PEAR gives the probability of being 3 parts in 10 billion that their remote-viewing results came from chance alone). However, the difficulty with this kind of remote-viewing trial is that the way it is undertaken is so subjective that it is hard to see that there can be any real confidence in these numbers.

One thing that has to be borne in mind is the cultural and social context in which the PEAR experiments were undertaken. These were times when conventional scientific thinking was being questioned by many in academia in a postmodern framework that seemed to suggest anything was possible. As the two main characters behind PEAR, Robert Jahn and Brenda Dunne, observe in their paper summarizing over two decades of work, "the primary scientific strands have been tightly interwoven with a number of philosophical, economical, political, cultural, personal and interpersonal fibers that have both constrained and enriched the course of research."

This statement does not in itself mean that science was put in second place to cultural and political aims, but bearing in mind those comments about the interwoven strands, we have to see this as an experimental program that took place in the same atmosphere as the infamous Sokal hoax. In 1995, physicist Alan Sokal got a paper published in a well-established academic journal called *Social Text*. In his paper, titled "Transgressing the Boundaries: Towards a Transformative Hermeneutics of Quantum Gravity," Sokal put together an intentionally meaningless collection of impressive-sounding words. His paper was a pure parody.

Sokal's intention was to demonstrate how writers in the humanities and the social sciences of the time were taking concepts from physics and using nothing more than a smoke screen of meaningless words and impressive-sounding jargon to totally distort the science and to pretend that they had in some way demonstrated that there were no objective truths of nature, but merely subjective interpretations based on the (bourgeois and limited) worldviews and the misguided culture of the scientists.

The genius of Sokal's parody is that he demonstrated that calculated garbage would be accepted by these academics because they hadn't the faintest idea of the realities behind what they were talking about. As Sokal has since pointed out, this wasn't just a case of putting down a few ivory tower professors spouting nonsense (thought that is quite an appealing thing to do, and probably a worthwhile exercise in its own right). It was also about defending science when it's under attack using spurious cultural arguments, whether they are nationalistic, gender-based, or religious.

I don't bring up Sokal's paper to suggest that PEAR's research was a fictional construct with no roots in reality—Sokal was, after all, criticizing theoreticians rather than practical people—but this paper and the reasons behind its being written should be borne in mind when we consider the context in which the PEAR work was analyzed and interpreted. There is more than a little reflection of Sokal in those words I quoted from the PEAR paper.

PEAR was dreamed up after some tentative positive results were obtained from an independent undergraduate project using a random event generator that depended on a noise diode, a mechanism that overloads an electronic component until there is a breakdown that, while not random in the same sense as a pure quantum event, is unpredictable enough to be considered a random source. The electronic device would effectively flip one way or the other, resulting in a series of binary digits that should have formed a neat normal (bell curve) distribution. The task of the subject in the experiment was to try to influence the reading to be different from expectation.

The main thrust of the PEAR work was to continue this experimental format on a much larger scale, making thousands of trials. We should acknowledge the steadfast and dogged work undertaken by the PEAR team. It isn't easy going against the grain of scientific thought, and ESP was considered pretty well outside the academic remit. There are many in the scientific community who are prepared to disparage psi phenomena without letting any consideration of the facts get in the way of their beliefs. When this happens, these dismissive individuals are behaving just as unscientifically as those who accept all psychic phenomena unquestioningly.

The controversial scientist Rupert Sheldrake has a good example of this approach in action. He was asked to take part in a TV show with zoologist and strident skeptic Richard Dawkins. To begin with the two seemed to find some common ground. Sheldrake and Dawkins agreed in front of the camera that controlled experiments were necessary to assess psi. The previous week Sheldrake had sent Dawkins copies of some papers published in scientific journals, and now he suggested that they discuss the evidence. According to Sheldrake, Dawkins "looked uneasy and said 'I don't want to discuss evidence.'" Sheldrake accused Dawkins of undertaking a low-grade debunking exercise. "It's not a low-grade debunking exercise," Dawkins replied, "it's a high-grade debunking exercise."

As it happens, I don't think the evidence Sheldrake presented, which was often anecdotal and not adequately controlled, stands up particularly well, but his experience shows how the starting point taken by many of the scientific community is a negative one, dismissing the possibility of psi without even taking the trouble to look at the evidence. As Jahn and Dunne comment, "Among our immediate faculty and administrative colleagues, the initial suspicion . . . has diffused over the years to a somewhat milder, albeit more widespread and generalized disparagement. In some cases this has been expressed by covert ridicule." It takes a certain amount of guts to continue for twenty years in such a climate.

Even if you accept the PEAR results at face value, which we will consider in a moment, there are some problems with them, but the way those values were reached has raised some question marks. In studying the long-term summaries published by the PEAR team, physicist Stanley Jeffers has pointed out two significant issues with the data.

One is that the PEAR team claimed that the effects produced when attempting to modify the output of the random electronic generator were "independent of the distance and the time at which the operator is expressing a particular intention that may be quite different from the time at which the device is operated." The independence from distance is relatively trivial with some potential mechanisms for psi, but that aspect of independence from the time when the device was operated is tricky.

The PEAR researchers appear to be saying that the effect registered on the generator can take place at a different time from when the subject tries to influence it. But as soon as you break the time connection between cause and effect, how is it possible to associate any specific change in the random generator output with a particular attempt to influence it? As Jeffers points out, "If true, it is not clear to this author how any causal relationship could be claimed between device operation and anyone's intention expressed any time." The simplest interpretation under such circumstances is not one of thoughts causing the changes, but rather that the fluctuations in output are independent of the experiment. Even if there was an influence with no fixed relationship in time, it could never be proved, rendering it inaccessible to science.

The second problem identified in the results by Jeffers concerns the so-called baseline behavior. As well as measuring the variations in output from the random generator when a subject was pushing for high or low values (leaving aside for the moment the alleged lack of a time connection between attempt and outcome), the experiment also involved periods of time when the subjects would attempt consciously not to influence the generator.

According to a paper on the data from the PEAR experiments, this baseline value was itself extraordinary. The expectation is that across a series of experiments the baseline value would wander around the average expected value. Sometimes it would be higher, sometimes lower. It should average out at the right value, but for some experiments it should be well away from the mean. But the suggestion in the paper was that the baseline value didn't stray enough. It apparently never moved outside the kind of levels you would expect it to exceed around 5 percent of the time. The suggestion was that this was a demonstration of a psi phenomenon called "baseline bind." Apparently, somehow, the subjects were forcing the neutral results to be more neutral than was expected.

When years later significantly more data was published, the concept of baseline bind seemed to have disappeared. This was just as well, because the baseline values actually did go outside the expected 5 percent level. In fact, from the plot of the values that PEAR produced, it looks like the baseline was gradually drifting over time, perhaps due to component aging in the apparatus, something that undermines the ability to combine data from different periods. The PEAR team disputes these suggestions, but it is a legitimate concern.

In retrospect, the problem with the PEAR project as a whole is that it concentrated on collecting evidence of tiny variations from random chance. It went out of its way to look for very small differences, rather than attempting to find anything definitive. The results were so marginal that had these experiments been coin tosses, they would have deviated from the expected 50:50 by only 1 in 1,000 tosses. With such tightly squeezed data, it is painfully possible for some aspect of the setup or analysis to cause the apparent anomaly that is being attributed to psi. Because the researchers produced a huge database of trials, in simple statistical terms the PEAR results were very impressive, with a better than 1-billion-to-1 chance of having a cause and not being purely random, but there is always the nagging suspicion that a phenomenon so marginal is, in fact, a kind of noise in the system.

Perhaps most damning for confidence in the PEAR results was the way that the results were opened to proper academic scrutiny. When any ordinary scientific experiment produces surprising results, the obvious next step is to attempt replication in other labs using the same experimental designs and protocols—and that is exactly what happened with PEAR. In 1996 laboratories in Freiburg and Giessen, Germany, joined with Princeton, which repeated its earlier runs to perform identical experiments.

To all intents and purposes these were experiments using the same approach, and employing identical methods for analyzing the data. The outcome came as a shock to those involved. Not one of the universities, not even the original Princeton labs, came up with any deviation from the expectations of random chance. The scientists taking part were unable to give any good reason why this should happen—but the inevitable conclusion is that the while the PEAR work was among the most properly scientific of any done on psi abilities, the results were so borderline that they were almost certainly attributable to some incidental cause.

It is very frustrating. At first glance, their results are very encouraging for those who support the reality of these abilities. And many person-years of effort went into gathering them. Yet there are real questions to be asked about the methods used at PEAR, and of the way the data has been interpreted. This kind of work, detecting tiny variations in statistical behavior, is simply not good enough to be a true demonstrator of psi abilities. We need clear, macro-scale evidence if we are ever to consider psi abilities to be scientifically proven.

I do not mean by this large-scale demonstrations where drama is more important than accurate measurement. But unless what we are testing for as psi makes sense as a detection of an actual ability, where we can clearly point to something happening rather than something to be deduced from a tiny deviation in a statistical expectation, the exercise is nothing more than playing with numbers.

The PEAR work was the absolute opposite of the flamboyance of a Uri Geller demonstration. As yet we have not met Geller, though his name has cropped up a good number of times, yet it is impossible to take a good look at psi phenomena without including this consummate showman. How should we treat the remarkable career of Uri Geller? After all, much of his early fame came from tests of his abilities performed in laboratories and published in respectable journals. Could this showman psychic really be demonstrating the remarkable powers that have been claimed for him in the past (and that he still claims for himself)?

10.

BENDING SPOONS

It is the evening of Friday, November 23, 1973, and British viewers are settling in front of their TVs to watch *Talk-In,* a chat show hosted by leading current-affairs anchorman David Dimbleby. The audience is about to get a shock. In silence, a hand holds up a bent key, nearly filling the screen. After a pause we hear Dimbleby's measured tones: "Tonight we meet the man who bent this door key, apparently just by stroking it."

Dimbleby introduces the audience to Uri Geller, about to take part in his first ever live TV broadcast. The scene is set by bestselling author and paranormal popularizer Lyall Watson. "Before we even see him do anything," a clearly excited Watson informs Dimbleby, "I think it's important to say one thing. That is that there are no tricks involved. . . . The first time I saw him everything was wasted for me because I was looking for the catch. There is no catch!"

With Watson's endorsement ringing in our ears, Dimbleby brings on Uri Geller. His guest claims to have two abilities. "What I do is telepathy . . . although I cannot sit here and know what you are thinking about me right now. It's very certain things that you really have to concentrate on. And I have the other power which I read that they call telekinesis. And that is moving or bending or breaking objects."

A tray of items is brought on and Uri Geller picks up an envelope in which a producer has placed a drawing. We are told that this is the first time Geller has seen the envelope. He asks that the producer concentrate on the image in the drawing. "It draws itself," Geller comments; "it doesn't just appear suddenly." He concentrates on the envelope for over a minute with occasional asides, an agonizingly long waiting time on live TV. Geller tells us he has picked up two things: a simple triangle and under the triangle a few straight lines, which after further concentration he decides could be a boat, scribbling his own version of the picture as he talks.

Another witness, mathematics professor John G. Taylor from London University, opens the sealed envelope and a second one inside it; he has some difficulty, fumbling under the pressure to perform on live TV. The envelopes contain the original picture, which Dimbleby tells us was drawn earlier that afternoon in a dressing room at the studio. A gasp comes from the audience, which turns into a roar of applause. Geller's drawing is nearly identical.

Next Geller is handed a watch, stopped at 6:20. The cameras are shown the unmoving second hand. A girl is asked to come up from the audience, and there is some confusion as panel members offer her their seats before she is given Dimbleby's chair and is asked to hold the watch. Despite several attempts, during which the watch is also passed to Lyall Watson to hold, it won't start. Finally, just as they are about to move on, Geller announces that the watch *has* started, and it is shown to a camera. To a big reaction, Lyall Watson suddenly announces that his own watch has stopped. This, Geller tells us, sometimes happens when he is around.

We then see some classic spoon (or more accurately fork) bending, where David Dimbleby holds one end of a fork and Geller rubs it near the tines, though the performer finishes the process off with the fork in his own hands only as the end of the fork bends and drops off. "Magicians can duplicate this!" Uri tells us. "But I want to see those magicians or those skeptical [*sic*] do it under controlled conditions where they cannot do it." At this point David Dimbleby wraps up the hugely successful broadcast with the information that the program's staff has already received calls from people who received the same image by telepathy or had watches started at home while watching the show.

Following the event, the BBC is said to have received "hundreds of calls and letters" relating how "in homes throughout Great Britain cutlery had been bent and timepieces, long defunct, restarted." With this performance, the career of the twentieth century's best-known stage psychic, Uri Geller, took off and headed for the stratosphere.

With his claims to be able to restart broken watches remotely, to bend spoons, to detect mineral deposits from the air, and to read minds, Geller is a walking psi laboratory. And he has repeatedly convinced scientists who have examined him at work that he is genuine. Yet many suspect that what Geller does is nothing more than stage magic dressed up as psychic ability to give it extra drama.

Geller was born in Tel Aviv, Israel, on December 20, 1946, but moved at age eleven with his mother to Nicosia, Cyprus, when his parents, Itzhaak Geller and Manzy Freud, divorced. Geller began his professional career as a nightclub magician in an act with a friend, Shipi Shtrang, who would later accompany him on many of his psychic ventures. Shtrang's constant presence as a possible feed and assistant is one of the major concerns that many observers have of Geller's performances. When working the nightclubs, Geller made no suggestion that he had true psi abilities; he was a magician pure and simple. Later, though, he would claim to have "first noticed his powers at the age of three, when he found he could tell his mother exactly how much she had won or lost when she came home from playing cards."

Geller sprang to fame following his European and American television appearances, particularly the Dimbleby show featured at the start of the chapter, receiving huge publicity in a book by Geller enthusiast Andrija Puharich, who described a whole mythos behind Geller's powers involving flying saucers and ancient gods. And that fame just kept growing as he was given his first scientific credibility by scientists Russell Targ and Harold Puthoff of the Stanford Research Institute.

Geller continued to gain more scientific coverage than anyone else in the field, including an article in one of the world's top science journals, *Nature*. It comments: "We present results of experiments suggesting the existence of one or more perceptual modalities through which individuals obtain information about their environment, although this information is not presented to any known sense. The literature and our observations lead us to conclude that such abilities can be studied under laboratory conditions." Strong acceptance indeed. Yet skeptics continue to assure us that Geller is simply a very clever stage performer, using tricks to fool the scientists. Can this be true when so many remain convinced to this day of Geller's genuine gifts?

Consider this enthusiastic commentary on Geller's performance on the British TV show by the initially skeptical mathematics professor John Taylor:

> No happening has ever been so dramatic as that shown by Uri Geller to millions of British viewers on that momentous evening. No methods known to science can explain his revelation of that drawing in the envelope. And similarly startling was the metal-bending. Here is a phenomenon associated with a form of matter thought to hide far fewer secrets than the mind and brain. This bending of metal is demonstrably reproducible, happening almost wherever Geller wills. Furthermore, it can apparently be transmitted to other places—even hundreds of miles away.

Other, less generous observers would suggest that while science could not explain Geller's copying of a drawing in an envelope, magical performers could easily duplicate it using readily available tricks. And the reproducibility of metal bending seemed, on closer observation, to be directly tied to the degree of freedom Geller had to manipulate the objects. (As for that ability to transmit metal bending remotely to places that could be hundreds of miles away, as we shall see this is more an aspect of human psychology than it is of telekinesis.) But there is little doubt that Uri Geller made a huge impact.

A typical description of a Uri Geller remote-viewing demonstration is as follows. Someone Geller does not know produces a number of drawings, selects one at random without Geller seeing which he or she has chosen, places it in an envelope, seals the envelope, and hands that to Geller. The participant is asked to concentrate on the image, and within five minutes, Geller has duplicated the image almost perfectly. That is stunning. He also performs similar telepathic feats where the participant is aware of the details of the image that is inside the envelope.

But here's where things get a little less simple. This is not a controlled laboratory experiment. What you have just read is not a detailed account supported by video evidence, but one person's recall of what happened. It sounds impressive, and the person relating the anecdote would be impressed. But it isn't a true analysis of fact yet. Now let's throw in a little more detail. In reality, the time period was closer to half an hour than to five minutes. During that time, when Geller asked the participant to concentrate, he encouraged him to close his eyes. For a couple of minutes, as the person who made the drawing sat with his eyes closed, Geller was totally unobserved. The envelope was quite thin and could have shown its contents if it was held up to the light. Or it could even have been opened and resealed.

There is, as Geller supporters would point out, no evidence in the real example I based that description on that Geller did cheat. But the important thing to notice is that the opportunities to cheat were there, despite the fact that the first account of what happened, the account that was likely to make it into the press, simply didn't bother to mention the detail that made the chance to cheat practical. When the same participant first wrapped his image in foil before putting it an envelope, and remained observant throughout, Geller tried for thirty minutes and got nowhere.

The telepathy version of his act is very similar. In the landmark BBC Dimbleby show, Geller amazed the audience by reproducing almost exactly the simple image of a sailboat drawn by a BBC employee and sealed inside two envelopes. On the show he asked the woman who had produced the original image, and who should have been the only person to know about it, to concentrate hard on the image. Geller than drew something almost identical, only differing in format by enclosing the boat in a curved frame, which he said was a mental TV screen that he used to visualize the image.

Remarkable indeed. Yet at no point was there any attempt to ensure that the image was not visible if the envelopes were held up to a very strong light, the kind of light that is to be found in abundance in a TV studio. And the most crucial piece of information is one that doesn't make it into the story. The image was not drawn immediately before the broadcast; it was produced several hours before the show went out, giving Geller or one of his associates plenty of time to sneak a quick look. Exactly how the envelope was protected from the moment it was sealed to the revelation on screen is unknown—the only assurance we have is that Geller himself tells the audience that he told the woman who drew the image not to let it out of her sight. And that is the key to knowing how Geller probably accessed the contents.

This is where, unfortunately, scientists who don't necessarily understand the mechanics of magic tricks are at a serious disadvantage. Mathematics professor John Taylor was present at a number of Geller performances, including his BBC TV debut, and makes it clear that he believed Geller was not cheating. After all, Taylor observed, the BBC assistant who drew the image was not known to Geller and would not have collaborated with him. "Too many people have drawn pictures for Geller's accurate guess for it to have been probable that all were in collusion with him," Taylor observes.

Taylor seems to have fallen for a classic example of misdirection. It would be interesting to know if the concept of collaboration was planted by Uri Geller—because a direct collaboration of this sort would be a relatively unusual mechanism for a stage magician to use in such an act. It is much more likely that he or a confederate (the ever-present Shipi Shtrang, for instance) either observed the picture while it was being drawn or managed to sneak a peek sometime between the original drawing being made and the performance. There was no need for a collaboration with the person who drew it, and no one was suggesting that it happened.

Sometimes the process used to perform a trick can be much less subtle. Geller appeared on the UK TV show *Noel's House Party,* hosted by Noel Edmonds. One feature of the show was a hidden-camera segment, when the cameras captured a celebrity in embarrassing circumstances. In this instance, Uri Geller is shown performing a one-to-one demonstration of telepathy with someone seated across a table from him. "What I do is real," Geller insists. "It is not magic and it is not a trick." Geller asks his female companion to draw something while he has his eyes covered and looks in the opposite direction; then he asks her to turn the pad over.

When Geller is allowed to look back, he asks his interviewer to project the drawing, seeing it as if it were on a TV screen—the familiar mechanism of his telepathic process. He tells her not to close her eyes, but he will close his. After a while he announces he has got something. Is it a little house? The interviewer is shocked. The picture Geller produces is very similar, even including the smoke coming out of the chimney.

What Geller does not point out is that this is by far the most common image someone will draw when asked to draw a simple picture. It is pretty well universal in Western culture. But in fact Geller was not relying on this. What is totally obvious in the video of the show is that partway through the drawing process Geller turns back briefly to face his interviewer. He still has his fingers over his eyes, but he is looking straight at her. She isn't concealing her artwork, which she draws with the pad flat on the table. It is hardly difficult for Geller to work out what she's drawing. She doesn't notice this, as she is concentrating on producing her drawing.

Perhaps the most fascinating version of Uri Geller's telepathy demonstration was given not by Geller himself but by James Randi. One of the important things this particular event demonstrates is that there are several different ways to work this particular trick. Randi's intention was to reproduce an effect Uri Geller had performed shortly after becoming famous. Geller appeared on Barbara Walters's TV show *Not for Women Only* and performed his usual telepathy act, managing to produce a reasonable copy of a drawing of two stick figures holding hands. Randi now intended to go on Walters's show and do everything that Geller had, but with no suggestion that he had any psi ability.

When Randi agreed to reproduce the demonstration, Walters's producer was determined to make trickery as difficult as possible. With the kind of control that has very rarely been applied to Uri Geller (and almost certainly wasn't on *his* Barbara Walters appearance), the drawing was kept in an envelope inside a book, well away from Randi, after it had been drawn.

Randi did not have a confederate who could try to get to the picture while Randi stayed well away. Walters (or her producer) literally had the book containing the image in her hand all the time from its being drawn to going on the set.

Alongside Randi onstage sat a pair of magicians, who admitted afterward that they thought he had no chance of an effective performance when they saw the controls that had been applied. It seemed the great skeptic would, for once, be hoisted on his own petard. Randi drew his version of the image while the envelope remained in the book. To Walters's obvious shock, when she displayed her image of a house with a stick figure alongside it, Randi turned his pad to face the audience to show essentially the same image. His person was inside the house, and he didn't have a sun in the picture, but otherwise what he had drawn was eerily close to the original.

"How do you do it?" Walters asked. "If you're going to debunk this man who everybody believes, you have to tell us how you do it." Randi ignored her, commenting that he was adding the sun to make his picture even better. "Come on," Walters pleaded, "tell us how you do it." But her request fell on deaf ears. What she had unconsciously identified is the biggest weakness in the frequently employed argument that magicians make the best people to uncover frauds in psi phenomena. Yes, they are great at spotting how a faker can produce an effect, but usually they won't say how they could duplicate it and hence allow us to actually spot the faker in action. They don't follow through. Until magicians decide that debunking fraudulent mystics is more important than losing a few of their trade secrets, they really fail to deliver all that they could.

In a video where he describes the encounter with Barbara Walters, Randi teasingly tells us that he has met Walters many times since and they "just don't talk about this." He has never had to give her an explanation . . . and she is still confounded. There may be hope that Randi will one day tell us how he worked his magic, but he hasn't yet. Because I'm not a magician and not sworn to secrecy, I would like to describe how Randi performed this trick, using one of the techniques that Geller very likely had in his armory alongside peeking through his fingers and getting a look at the picture when it is unattended.

It's just possible that Randi used the statistical evidence and went for a house because that's the most likely thing for someone to draw (and he knew that Walters wouldn't draw people alone, because she was unlikely to repeat what had just been demonstrated on tape in her experience with Geller). However, I think there are a number of clues that Randi gives out that provide a pointer to the way he really did it. And I refer you to perhaps the most famous of all quotes from Sir Arthur Conan Doyle's Sherlock Holmes stories, when Holmes says in *The Sign of Four*: "How often have I said to you that when you have eliminated the impossible, whatever remains, however improbable, must be the truth?"

It is more than a little ironic that Conan Doyle's logic should be used to uncover this kind of trickery, as the author was notoriously bad at spotting fakes. He famously publicized a series of photographs of fairies taken by two schoolgirls in the north of England. Although these pictures blatantly featured paper cutouts, Doyle doggedly continued in his belief that they were real fairies until his death. But Holmes seems to have had a better brain than his creator. The Sherlock Holmes argument tells us that if it was absolutely impossible for Randi to have seen Barbara Walters's drawing before he made his, unless he relied on good guesswork he must have made his drawing *after* he (and the audience) saw her original.

What we see when watching the show is Randi apparently drawing his copy of the picture on a pad using a ballpoint pen before Walters reveals her picture. It is possible, even using the technique I'm going to suggest, that Randi did do a little drawing at that point in the proceedings—if so, what he produced was probably a basic box, which he could adapt later for whatever was needed. It's equally possible that he didn't draw anything, but merely moved the pen to make it look as if he were putting something on paper.

When Walters shows her picture to the camera, and all eyes are on the image, Randi is holding his pad in front of him, with the writing surface facing his body. He isn't holding the pen, so he can't be drawing anything, right? Except there is an old magician's trick of fixing a pencil lead under the fingernail, and using that to draw something unseen, concealed behind the pad. The same thing can be done with the end cut off a ballpoint pen refill. What you get, in effect, is a finger end that draws like a pen.

This is, I'm convinced, how Randi performed the trick, adding details to the image while Walters was displaying her picture to the camera. He couldn't look at his own picture much as he did so or it would have given the game away—and this would explain why his stick figure ended up on top of (or as he put it, "in") the house rather than alongside it. One clue that Randi may have given when he described what happened is that he made a big thing at the time of adding the sun to make the picture more like the original—emphasizing, perhaps, how he worked by adding drawings after the event. Also, he would later stress in a video where he discusses the event that he used a ballpoint pen, while Geller used a big marker, making it harder to duplicate this technique (in fact, Geller drew his image in plain sight, so he couldn't use this technique).

The same kind of gap between the way an event is perceived and described and what actually happened can often be found with the remarkable demonstrations that Geller has made his own: telekinesis taken to a new and dramatic high—spoon, fork, and key bending. The typical description of a Geller event in a crowded room makes it sound as if items were bending of their own accord all over the place. Yet try to pin down precisely what happened, and you will find that the initial bending never took place in plain view. (It is easy enough to fake a bend appearing to increase in magnitude as an object is held in the hand by the way the object is moved through the fingers, but the initial bend has to take place somewhere.)

During his demonstrations Geller hardly ever is still. He may have to go across the room on a perfectly innocent activity, or to shake hands with someone, or to have his photo taken. Similarly, he often comes into close contact several times with an assistant, who can take an object, provide the bend out of sight, then pass the item back. Under detailed analysis, the initial bend with an object Geller has not had an opportunity to handle earlier never takes place in front of the audience's eyes. Where, as happened in the Dimbleby program, the bending seems to take place without the tableware leaving our view, the likelihood is that the cutlery was pretreated to make it easy to break with the small amounts of force that Geller can easily apply in the many manipulations he undertakes.

It might seem impossible for the fork Geller bent to be pretreated in the sort of controlled circumstances that took place in the BBC TV show, but in fact, on closer observation, everything was not what it seemed. On the Dimbleby show, Geller picked the item he was going to bend from a tray of cutlery. Of course, we assume, he didn't have access to it beforehand to be able to prepare and prebend some of the items. But it turns out that when later questioned, the director and the producer of the show had to admit that Geller had asked for the tray of cutlery to be placed in his dressing room before the show—and they went along with his request.

No doubt Uri Geller will have said that he needed to do this to "get in tune with the metal" or some such excuse, but the fact remains that the item he would later bend was not kept securely away from Geller and his associates, but was in his dressing room before the show—and this was never mentioned on air. How could those putting the program together be so naive as to allow an obvious opportunity to cheat? They thought that they had things under control by having someone guard the tray to make sure Geller didn't interfere with the cutlery. But when the producer went into the dressing room at one point before the show, he found that Geller had the tray of cutlery all to himself. He had sent the guard out of the room on an errand. Job done.

As for Geller's watch-starting feats, many stopped watches will start briefly if given a vigorous shaking. Just take a look at the description of Geller's watch-starting trick on the Dimbleby show at the beginning of the chapter. We don't see the watch starting under close observation. What we see is Geller taking the watch, then providing some excellent misdirection by making arrangements for a girl from the audience to come up onstage, causing a fair amount of confusion among the panel. During that period, no one is looking at what Geller is doing to the watch—in fact, most of the time the cameras are off him, concentrating on the antics with the audience member.

This gives him ample chance to give the watch a number of vigorous shakes to try to get it going or, as he sometimes does in his act, to adjust the hands of the watch so that it appears to have jumped to a different time. Crucially, on the TV show, we never see the watch between Geller handling it and it being revealed working in the woman's hand. As soon as he gives it to her he closes her hand over the watch. In this particular instance the watch didn't start until there had been several other distractions, but the method of misdirection is very clear.

Ironically, some would argue that Geller's initial failure to start the watch makes it more likely that he is genuine—surely, they say, a faker would get it right every time. But the fact is that watch starting is not an exact science—it depends on an element of luck. Not every stopped watch will restart with a vigorous shaking, but some will. Geller's regular failures seem to be either with watches that won't start on being shaken, or in circumstances where the control conditions are sufficiently good that he and his confederates simply don't have the opportunity to cheat.

Another aspect of Uri Geller's apparent ability to influence watches and clocks is that his ability seems to extend beyond the studio. It's not just watches he can handle onstage that miraculously restart; many viewers or listeners have called in during and after the many shows that Geller has appeared on to describe how their own watches and clocks have spontaneously started or stopped. As John Taylor commented, he seems to be able to extend his ability beyond the studio for hundreds of miles. The effect is even broader than simply influencing watches and clocks—anything can happen, from pictures falling off the wall to plates being spontaneously broken.

Clearly Geller can't be rushing round the houses of all these viewers and listeners to manipulate their possessions. This can't be the same kind of misdirection and interference as takes place onstage. Could it be that there is a real psi phenomenon in action here? A useful indicator of what is at work in these cases comes from a phone-in radio show featuring another mental performer, James Pyczynski. As he spoke on air, attempting to send out his influence, a whole string of people called in to describe the weird effects he was producing in their homes.

Clocks started, stopped, and sped up. Animals began to behave strangely. Mirrors cracked and a lightbulb exploded. Items around the house bent and broke. All until it was revealed that Pyczynski was James Randi's assistant and had no strange abilities whatsoever. There was no connection between his appearing on the radio show and the reported events.

The fact is that things happen all the time. I just heard a strange cracking noise from the kitchen. Well, houses make noises. We all occasionally hear strange noises, have pictures fall off the wall, have clocks stop or start, have pets that behave strangely. Usually when this happens we briefly comment on it and then forget all about it. There is no reason to attribute what we have experienced to a particular cause—it just happened. But should it take place at the same time as a performer who claims special powers is appearing on the TV or radio, it's easy enough to forge a false link between the two and believe, genuinely but without good reason, that the performer is causing the phenomenon.

Other "strange happenings" go unnoticed until you look for them. That broken mirror or cracked photo of your mother-in-law could well have been like that for weeks. But it is only when you are on the lookout for anything connected with a psychic performance being broadcast that you notice it. And other such reports are likely to be simply untrue. Human beings are strange creatures and do quite often make things up simply to go along with the crowd or to get noticed. But even taking just the events that actually do happen at the time, it would be very odd indeed if nothing whatsoever occurred during a show broadcast to millions.

Just taking the example of a clock that suddenly stopped during a TV show—let's assume that most people have a clock stop once a year (a conservative estimate, certainly in the 1970s, when clocks were more common in houses). And let's say that a million households watched a particular show. The show might last an hour—but many people will associate an event with the show if it happens half an hour either side of it. So we have a two-hour window in which spooky things can take place and be blamed on the mental performer. How many of those million households would we expect to have a clock stop with that length of window? There are 4,383 such windows in a year. So around 228 viewing households should have a clock stop to their surprise. That's a lot of stopped clocks.

Part of the problem here is the "unlikely things will usually happen if you have a large enough sample" effect. It is common to use winning a major lottery as an example of something that is so unlikely that it will never, ever happen. When we talk about something that is very unlikely we say, "There's more chance of winning the lottery." Yet the fact is, week after week around the world, people do win lotteries. It's true that the chances of any individual winning are very small. But the chances that someone—anyone—will win are very good. Similarly, the chances that your clock will stop while a psychic performer is on the TV are low, but it would be very strange indeed if no one's clock stopped.

Interestingly, when Geller appeared on the *Tonight* program with Johnny Carson, who was himself an ex-magician and who had James Randi to advise him of suitable precautions, the show's management made sure that Geller had no opportunity to interfere with the props beforehand. On a painfully slow and uneventful show, Geller was unable to do any of his usual tricks.

Everyone, even Geller's supporters, would agree that he is sometimes caught cheating (though it is often passed over by the media when it happens). *Daily News* reporter Donald Singleton, who spent several days with Geller, discovered at one point that Geller had a napkin at his place at a table concealing a pre-bent fork. The obvious assumption would be that this was a clear example of cheating—a prop that he had planted for use later. Geller claimed that it had simply happened of its own accord due to his mystical influence. Geller has also been caught on camera bending a key in a hidden-camera show on the Italian channel RAI, with clear hand movements producing the initial bend in a key that would later be gradually revealed as if it were happening at the time.

In another demonstration, on German TV, Geller performs the remarkable feat of breaking the end off a thick metal ladle. Everyone is very impressed. Geller chose this item from a tray of around thirty metal objects provided by a physicist who was present as an expert witness—there was, it seems, no way that this ladle could have been doctored. And yet . . . although not allowed to say anything on camera, the physicist pointed out after the event that while he had indeed provided the cutlery for the test, there was something strange. It seems there was just one item on the tray that he did not provide, and that no one involved in the TV show knew anything about. That one item was the ladle. If there is no chance to doctor the provided samples, the answer seems to be to provide your own and brazen it out.

Those who believe in Geller's ability but accept that he does sometimes cheat will say that this is a common behavior of genuine psychics. Such people have a natural urge to please their audience, but their psi powers are changeable and don't always come on command. So when necessary they will cheat to keep things going. But, argue the supporters, this doesn't prove that someone like Geller is always cheating. This seems, however, a feeble defense. While it is, indeed, not proof, it is a very strong indication that psychics can't be trusted. The fact that a dog only sometimes bites people does not make it a well-behaved dog.

Because of the coverage in *Nature,* Geller's testing at the Stanford Research Institute is particularly important to understand if we are to decide whether or not he is genuine. On Geller's website, the SRI experiments are given pride of place in a section labeled "Science," where we read, "These important controlled experiments were published as a scientific paper in the prestigious British journal *Nature.*" With the kinds of claims being made it is important to establish three things:

1. Did *Nature* really publish this paper? Most people won't check up on such a claim but will just take it as truth.
2. Were the experiments really controlled? As we have already seen, just because an experiment was performed by scientists does not mean that it had mechanisms in place that would effectively prevent cheating.
3. Were the experiments a success? Again, many who hear this claim will take it at face value without finding out what the actual results of the experiments were.

There is no doubt about *Nature* having published the paper—it is still available in the journal's archive—but what Geller's website, and many other sources, carefully omit is any reference to the accompanying editorial. In this piece, the editors made it plain that the paper was being published with significant reservations. They were not happy with the quality of the paper or the experiments as described. The editorial tells its readers that details of the way the experiments were carried out were "disconcertingly vague," and would be unlikely to be accepted by a psychological journal.

The editor decided to go ahead with publication because, despite its shortcomings, it was a paper presented as a scientific document by two qualified scientists (Targ and Puthoff were both respected laser physicists) "writing from a major research establishment apparently with the unqualified backing of the research institute itself." The feeling was that lessons could be learned and that publishing the paper would encourage a debate on the controversy, which would be better than science ignoring the ever more strident claims in the media of Geller's powers. With these provisos, the paper has a very different significance from that implied when it is taken in isolation.

Geller's website describes the experiments as controlled—but were they really? This is a crucial part of the analysis. As we have seen time and again, left to their own devices it is easy for those who are faking psi abilities to do so if measures aren't in place to strictly control what is happening.

There were two important types of test in the SRI trials of Uri Geller. The first covered telepathy and remote viewing, usually involving reproducing a picture while positioned in a different room—quite similar to the Derren Brown demonstration described at the start of chapter 6. The remainder involved a six-sided die in a metal box. Geller was tasked with deciding which face of the die was upward after the closed box had been shaken.

Good controls would mean that in the picture-drawing experiment there should be no way to communicate using normal means between the room where the original was located and the place where Geller was receiving his images. There should also be no one who could possibly be working for Geller able to see the original pictures Geller was attempting to reproduce—either inside the room where the pictures were, or able to see into that room from elsewhere. Although perfect control of communications would make this restriction unnecessary, in practice it is almost impossible to be absolutely sure that no communication is possible, so it is good practice to avoid the possibility of using an accomplice to pass on information.

In the die experiment, there ought to have been no opportunity for Geller to tamper with the die or to substitute a loaded version. The box should have been totally opaque with no cracks, and it should have been sealed in such a way that only the experimenters could open it. Ideally, Geller would never have touched the box—if Geller claimed that this was necessary (which seems unlikely for remote viewing, which by definition is supposed to work at a distance), then the box should have stayed in one of the experimenter's hands, or in a controlled environment like a vise, while Geller was attempting to detect the value.

With these or similar precautions in place, plus detailed video coverage, the experiment would indeed have been controlled, just as Geller claims—and as Targ and Puthoff say was the case in their paper. But one of the complaints that the editorial introduction in *Nature* makes is that the details given of the controls applied were "uncomfortably vague." From what has emerged since, this was with good reason. From an edited video made during the SRI experiments and as a result of information gathered from others present during the experiments, it seems there were huge flaws in the controls that made it entirely possible for Geller to achieve what he did without resorting to psi.

In the *Nature* paper, we are told that for the image-reproduction experiments, Geller was placed in various locations, but the most impressive was a double-walled steel room with a one-way audio monitor that allowed sound out but not in. Targ and Puthoff write, "in our detailed examination of the shielded room and the protocol used in these experiments, no sensory leakage has been found." Yet according to other observers there were at least three problems with Geller's isolation in this location.

First, the room was not soundproof—sounds from outside could carry into the room. This was inevitable because there was a hole in the wall between the "sealed" room and the location outside from which the pictures were transmitted. This 3.5-inch hole had some gauze pushed into it in an attempt to block it, but such a feeble precaution would not prevent sounds from passing through, and on at least one occasion the gauze was found to have fallen out. This hole could easily have been opened for visual communication had Geller had a confederate in the outer room.

Second, the room in which Geller was confined was not tested for radio isolation. Stage magicians often use small earpieces to receive information about members of the audience. A good magician would have no problem ensuring that he could take this into the sealed room without its being found. (Houdini could secrete a key on his person despite being stripped naked and given a medical probing.) If radio communication was possible and there was some kind of helper for Geller on the outside, then the room would not hide many secrets.

Other locations used in the experiment were even less well protected. One was a Faraday cage, which admittedly should in principle prevent electromagnetic communications. A Faraday cage relies on the fact that being surrounded by a metal box (even if it has holes in it) prevents electrical charges being transferred to the inside of the cage, making radio communication nearly impossible. But the cage had large openings that could be seen out of, making it hopeless as a way of visually isolating Geller from a confederate.

The very fact that Targ and Puthoff didn't develop a good control protocol and stick with it is unfortunately typical of the amateurish nature of so many psi tests. Why keep switching locations? It is as if the experiment were run by children who didn't have the patience to stick with a particular way of doing things. By constantly presenting new opportunities to get around their controls they just made cheating easier, should Geller desire to do so.

And the third problem? What is never made clear in the *Nature* paper or anywhere in Targ and Puthoff's descriptions of the experiments is that Geller's longtime stage co-artist Shipi Shtrang was present in the outer room during the experiments. One psychologist observing the tests commented that Shtrang was "constantly under foot." With his professional expertise, Shtrang would have no problem either observing aspects of the images or communicating them using sign language or radio communication, or even on a slip of paper conveyed through that hole in the wall. (It is interesting that the time the gauze was recorded as falling out corresponded to the only test in which Geller made an exact reproduction of an image, a bunch of grapes, down to the number of grapes in the drawing, something that would most easily be communicated on paper.)

As for the die experiment, this sounds highly impressive taken at face value. In the *Nature* paper, the authors tell us that "a double-blind experiment was performed in which a single 3/4-inch die was placed in a 3×4×5-inch steel box. The box was then vigorously shaken by one of the experimenters and placed on the table, a technique found in control runs to produce a distribution of die faces that does not differ significantly from chance distribution. The orientation of the die within the box was unknown to the experimenters at that time. Geller would then write down which die face was uppermost."

That sounds great. It is excellent that there were control runs to check the fairness of the die, and that the experiment was double-blind, so the experimenters' knowledge of the target value could not in some way be leaked to Uri Geller. But the authors do not point out two critical facts. From the description, it sounds as if the box were isolated on a table while Geller performed his remote viewing from a distance. In fact, not only was he in the same room, he was allowed to handle the box for quite considerable periods of time before making his prediction. Couple this with the second fact, that the box itself was not in any way sealed or locked, making it easy for a consummate professional to sneak a peek when not observed, and suddenly the die experiment transforms from a true feat of remote viewing to a basic conjuring trick.

The controls, then, were practically nonexistent. And what about the success rates? With the die, Geller had a fantastic hit rate. Out of ten trials he could not come up with an answer in two cases, but in the other eight he was correct each time, a million-to-one coincidence. But this is hardly surprising if he managed to peek—and the fact that there were some trials in which he passed on producing any answer (a common feature of Geller performances) is highly suggestive of there being occasions when it proved impossible to cheat safely.

In the image-transmission experiments, he achieved a number of successes, but again there were images that he refused to attempt. The only totally clear reproduction he made was the bunch of grapes. With other images he came up with drawings that were different from the original but reminiscent of it, in a way that could occur if someone was trying to indicate to Geller with gestures what was in the picture.

Perhaps the most interesting of the images transmitted is a picture that featured a devil with a pitchfork. Geller produced a plethora of images, covering much of a sheet of paper, practically all of which had nothing to do with the original. Why would he need to do this if he truly was capable of remote viewing or telepathy? But one part of the drawing was a pair of hastily scrawled pitchforks. These are clearly drawn differently from the rest of the images—much more sketchily, and even crossing over one of the other images. The way those pitchforks are drawn is very suggestive of the technique I suggested that Randi used in reproducing a picture (see page 220), adding the image after the original was revealed, using a pencil lead or the end of a ballpoint pen refill under the fingernail to draw blind while not looking at the paper. These pitchforks look distinctly as if they have been added in this fashion.

Interestingly, when Geller was given a different test with better controls, where no one knew what images were in a set of one hundred envelopes he passed on every single envelope, despite trying over three days to view what was inside them. This is surely a very important result, yet it isn't something you will find mentioned on Geller's website or in enthusiastic books on Geller's powers. Nor is it mentioned that on the one occasion the experimenters allowed the loosening of these conditions, Geller could suddenly see what was in the envelopes. (This doesn't feature in the *Nature* paper either.) An SRI psychologist, Charles Rebert, commented, "I was there, and I'm convinced that he cheated."

Inevitably some of the analysis of Geller's performances is supposition, but by no means all of it. Apollo astronaut Edgar Mitchell, a Geller enthusiast, was present during the trials. While the inclusion of celebrities seems more suited to a stage performance than to a scientific experiment, Mitchell's presence was useful, as he was able to give an estimation of the quality of the trials that rings true, especially bearing in mind that he was no skeptic, but instead an outspoken supporter of the existence of psi abilities: "[Targ and Puthoff] were so eager to keep [Geller] around that they worked themselves into a box by meeting his every whim and if he threatened to walk off they would relent and do what he wanted. Of course, they lost control of the situation and it just got worse and worse and worse." So this was the famous "controlled" scientific SRI experiment. Geller's every whim was met—whims that often seemed suspiciously like opportunities to cheat.

Even though it lacked the controls that should be applied in a lab, a more effective judgment was made when Geller demonstrated some of his abilities in the office of *Time* magazine in 1973. Unknown to him, one of the journalists present was the magician James Randi. Geller performed feats of two of his hallmark abilities—telepathy and metal-bending telekinesis.

The telepathy came in two parts. In the first, Geller attempted to reproduce drawings made by the others. Randi, as the sender, held his pencil so it was hidden behind his pad as he drew and Geller could pick up nothing. Others allowed the ends of their pencils to be visible, and Geller, peeking through his finger-covered eyes, managed a fair attempt at reproducing the images that seemed to depend on the classic trick of watching the movement of the end of the pencil and deducing what is being drawn.

In the second attempt Geller wrote down first a capital city and then a number between 1 and 10. Those present had to try to pick up what he had written as he mentally broadcast the results. Someone mentioned London—it turned out Geller had written both Paris and London, but had lightly crossed out London. This way he had two possible hits on the most likely international capitals to be guessed. It should come as no surprise at this point that the number he had written down was 7—by far the most likely number to be guessed between 1 and 10. This was the sort of "magic" that students used to do when I was at school and that didn't seem very impressive back then.

Geller then bent a key and a fork. Randi and other witnesses claim quite simply that he physically bent these—in the case of the key by simply pressing it against a desk. Because he used other actions to redirect the attention, some people did not notice, but Randi kept his gaze on the key at all times. The key was later found to be even more bent. A miracle? Perhaps not when you consider that, in Randi's words, "Uri took the key and rushed out of the office and down the hall with it." This kind of activity seems very common in Geller's performances, where he (or a confederate) often moves briefly out of sight during a performance, holding the items to be bent.

What is particularly interesting about this *Time* demonstration is that it was also written up by Andrija Puharich, the man who brought Geller to the United States. Puharich's version is very different from Randi's. In Puharich's description in his book *Uri*, he and Geller were well aware that there were magicians present, and despite Uri's successes, these magicians were constantly sniping, saying they could produce the same effects by trickery. While it's indubitably true that the tricks could have been duplicated, this version of the story differs not only from Randi's statement but also from confirmation by the *Time* staff present who had no ax to grind.

This contradiction is typical of many accounts of Geller's work. Many eyewitnesses are convinced they have seen one thing, when in fact something totally different happened. There is no suggestion that Puharich was anything other than a true believer in Geller's abilities, but his desire to see what he wanted to see meant that his account of the event was totally at odds with the more objective observers from *Time*. Bear in mind that Puharich hoped to make a good amount of money from a book in which he described Geller's abilities. He needed wonders to take place.

Puharich has been responsible for some of the more extreme aspects of the Geller story, suggesting in his book that the performer gets his power from a flying saucer. Perhaps because of the reduction in credibility that goes along with this story, this is a suggestion that Geller himself does not now repeat (though I am not aware that he has ever directly contradicted Puharich). Allegedly Puharich witnessed Geller entering a UFO—a disk-shaped metal construction with a blue flashing light on top "they found in the desert." He claims that there was film of this incident but "unhappily the film cartridge containing the essential record of this event was lost, having dematerialized within a few minutes." As indeed it might.

Randi documents a fascinating example of how judgment is colored by seeing what you want to see. He visited a Geller enthusiast, Dr. John Halsted of London University, who, along with world-class physicist David Bohm, had witnessed Geller at work and had been astounded by the way a spoon Geller broke in two did not have the stress fracture markings that he expected it would show if the spoon had been fatigued by repeated bending back and forth.

Randi asked Halsted to take a spoon from the canteen and slip it in his pocket. They then returned to Halsted's office, where Halsted took a phone call and then Randi made the spoon break by (apparently) simply twiddling it in his fingers. Halsted discovered that this spoon also did not have the fatigue marks. Randi does not disclose how he performed this trick (although as always he is very clear that it *was* a trick), though he confirmed to Halsted that he did not bend it while Halsted was on the phone. What seems much more likely is that Randi earlier put another identical spoon from the canteen in his pocket, which he could work on as much as he liked before substituting it for Halsted's spoon.

What is most interesting here is Halsted's reaction to Randi's trick. At the time he believed that Randi was a reporter. When Randi performed the trick, Halsted dropped the spoon into the bin, commenting that he wouldn't want it confused with the "real" experiments. There seemed to be no surprise that a mere reporter could duplicate Geller's trick, nor did the suggestion occur to him that this meant Geller's methods had to be questioned. Halsted had two totally different mind-sets for his dealings with Geller and with the normal world, something Geller fosters with his whole persona and patter. This mind-set problem that believers in his abilities have is as much a part of what is happening as any physical demonstration.

Geller also performed apparent feats of telepathy when numbers and words were written on a blackboard in his absence, or he described items that members of the audience had about their person. These apparently impressive demonstrations were explained with the help of friends of Geller in a remarkably frank article in the Israeli weekly magazine *Haolam Hazeh*.

These demonstrations relied on Geller reproducing information that he couldn't possibly have witnessed himself because he was not present. However, this is where Shipi Shtrang and the other members of his entourage came in. Geller always insisted that these people had good seats in the auditorium. As we have already seen with his exploits at SRI, Shtrang particularly was often present and "under foot" while testing was under way.

The article quotes Shtrang's sister Hannah, who described how Geller and Shtrang met when Geller was a counselor at a summer camp that Shtrang attended. A friendship began then that had lasted ever since—and they proved to have a mutual interest in stage magic that eventually developed into Geller's act. According to Hannah, Uri Geller's mind-reading tricks depended on Shtrang sitting in the audience as a plant. The two had agreed on a series of signals that Shtrang would use to pass information to Geller. Hannah, Geller's chauffeur, and other friends all admitted to also taking on this role during different shows.

So, for instance, when a number was written on a blackboard in Geller's absence, Shtrang would use a simple code, linking a number to gestures like touching a particular eye, licking the lips, touching an ear, and so forth. Whenever there has been a suggestion that the blackboard trick be performed in such a way that the audience (and hence Geller's helper) can't see the information written on it, so only the person who writes on the board knows what is there, the trick has been canceled or has failed.

Related tricks were often performed using a "psychological force" where the performer makes use of the very limited range of guesses an audience member is likely to make. Geller, for example, often performed a mental trick of picking up a world capital that an audience member is thinking of—but as we saw with the London/Paris example, in practice there were rarely more than two or three places selected. As far as I am aware, Geller hasn't performed the trick I am about to describe, but it demonstrates how the structure of an illusion can be used to draw attention away from the point at which the performer misleads the audience.

Imagine you want to fake the ability to communicate mentally with complete strangers. This requires no stooges or plants. You ask all the audience to sit back, clear their minds, and think of a color. They need to concentrate hard on this color, but mustn't say anything aloud. You peer at the audience, clearly mentally straining to pick up a transmission. After a while you ask three audience members, widely distributed across the audience, to stand up. These three, you say, are particularly strong telepaths. They smile, looking rather pleased with themselves.

"Okay," you say to the three people who are standing up. "When I mention the color you are thinking of, sit down. I'm getting . . . just really focus, your minds . . . I'm getting blue, red, and green." All three sit down. The audience is impressed. It appears that you have correctly identified the three colors they were thinking of, one for each individual. But in practice you didn't have to. Most people will be thinking of one of those three colors, but it's quite possible they all were thinking of the same one—blue, say. However, your statement was phrased in such a way that any combination of the three colors among the participants would appear to produce a perfect guess. No mental powers required.

Of course, sometimes you will get it wrong because someone has cleverly thought of, say, cerise. No problem: a performer with Geller's excellent ability to think on his feet would just say that he had seen that color as red, or if there is no way to link the color to any of his suggestions, then he would simply point out how subjective our perception of colors is. Under those circumstances, he would say that what they saw as, say, brown, he *felt* as green. I am not saying that Uri Geller has ever used this specific trick, but there is good evidence that the underlying technique being used here, the psychological force, is part of his toolbag.

As for the apparently remarkable ability that Geller sometimes demonstrates of knowing some of the possessions members of the audience have on their person, according to the whistle-blowing article, Geller's assistants are instructed to keep watch on people as they arrive, when they interact with the box office, when they make purchases, and when they settle into their seat. The confederates note any objects that come into view, which Geller can then describe, to their owner's amazement.

Hannah also describes how her brother and Geller spend hours practicing reproducing images having got only the briefest glimpse of them, and then signaling the content of an image with the minimal amount of contact. This article, which never seems to have made it into the awareness of the world's press even when published in translation, is the clearest indication that Geller's tricks are just that. When a magician shows that he can reproduce a trick it merely demonstrates that it is *possible* that Geller is cheating. When close confederates describe how he does his tricks, then there can be little doubt remaining.

In the case of Uri Geller we have someone who has set out to make a career out of apparent psi abilities, and there are plenty of other similar professionals who haven't achieved Geller's fame (or notoriety). However, before leaving the dubious side of parapsychology research we ought to consider another kind of manipulation of those genuinely trying to study psi phenomena, which could be motivated by either the desire for attention or sheer malevolence.

There are a number of examples of this happening, but perhaps the most striking case is the fooling of British mathematics professor John G. Taylor. Taylor enjoyed media exposure and had an interest in the extremes of science, so it doesn't seem surprising that he set out to test for the existence of Geller-like abilities in ordinary young people after his appearance as an observer on the Dimbleby show. Taylor documented his own experiments and theories on psi in his book *Superminds*.

Taylor spends a considerable amount of the book describing Uri Geller's performances, with a worrying ability to miss the point. Commenting on Geller's ability to forecast the die throws in the metal box at SRI, he comments on how unlikely it was that Geller was able to use a special die with radio transmitters that showed which side was up. This was technologically unlikely at best—but more to the point, it was irrelevant. All it took to cheat was to sneak a peek at the die; there was no need for high-tech trickery. Similarly, Taylor claims that the SRI telepathy and remote-viewing experiments could be fraudulent only if there was "gross collusion" between Geller and the researchers, but he fails to point out (in fact may even be unaware) that Geller had his assistant present to help with any collusion that was needed.

The most original aspect of Taylor's work in this field is his work with young metal benders in the UK. As he writes, "Geller's three consecutive British appearances, on the Jimmy Young, Dimbleby and Blue Peter programmes, set off a rash of spoon-bending all over England. . . . Scores of people suddenly discovered themselves to be metal-benders and hundreds claimed they could start and stop watches at will. The initial burst of metal-bending was followed by a steady increase which still continues."

Taylor goes on to relate a string of anecdotal examples of metal bending that took place after Geller's performances. Of themselves these stories are fascinating but have no scientific value, as Taylor himself must have been aware. But, to his credit, Taylor was not prepared to rely solely on anecdotes and set out to provide a controlled set of experiments that would allow him to monitor exactly what was happening with a number of children who claimed to have remarkable metal-bending abilities.

Taylor notes that a surprisingly high proportion of those claiming such abilities were children. It is worth taking a quick look at his explanations for this, as they help give a picture of the mind-set with which he was setting out to test these children. Perhaps, he suggests, the children have more chance to practice their special abilities because they have more spare time than adults. Or perhaps there is some aspect of metal bending that has evolutionary survival value for children, but not for adults, hence the ability fades away as we get older. What he fails to put forward is the clear and obvious possibility; perhaps it is because children are more likely to make things up and to accept fantasy elements in their lives, and even to manipulate the world to better fit those fantasies.

Taylor began his experimental work by sensibly moving away from cutlery—this is fine for a TV show but is hardly a standardized instrument for use in a scientific experiment. He wanted a simple object to bend, and one where he controlled the origin, rather than using a spoon or a fork that could have prepared in advance by repeatedly bending it to weaken it.

His chosen test materials were strips of copper and aluminum around 10 centimeters (4 inches) long and 0.6 centimeters (0.34 inches) across. These were the central experimental tools of his telekinesis experiments, though Taylor also tried a huge range of materials—plastic, glass, carbon, and wood, as well as other metals, some with the softness of lead and tin, others with the solidity of iron and tungsten. He found that on the whole anything could be distorted or broken, with the exception of glass.

In some ways, Taylor was very careful in his research—it is interesting and perhaps typical of those trying to observe such phenomena from a physical sciences background that he put much more effort into establishing whether various physical phenomena were involved in a particular activity than he did into ensuring that there was no human interference. So, for example, he describes how he looked into whether external electromagnetic radiation played a part in the bending.

To do this, he tried placing various strips inside different shields. Some were in two different types of tube of fine mesh, which acted as Faraday cages to block out electromagnetic radiation. Others were in a quartz tube, which lets light through but blocks most other forms of electromagnetism. Taylor observed that his test subjects seemed to be blocked by the tubes in the laboratory, but "two of the best metal-benders, after a week at home with all three types of tube, returned them with the aluminium strips fantastically distorted and none of the seals had been broken."

Notice what is being described here. Taylor is going to considerable lengths to see if electromagnetic radiation is twisting these strips—a very unlikely possibility at best—yet he seems oblivious of the fact that he was letting his subjects take the strips home for a week to work on them at their leisure and out of sight, relying on the children's honesty and the seals he had placed on the tubes to ensure that there was no tampering.

In another description of metal bending where the tubes weren't involved and the metal was directly accessible to the hand, Taylor says, "Another subject, who had been *left with* a strip of copper, proceeded to break it into pieces by producing softening, even going so far as being able to pull off a small piece" (my italics). I think that I, with no claimed metal-bending abilities, could break a thin strip of copper if I was left with it.

Taylor goes on to describe all kinds of subtle and unlikely causes for the ability of his his "super children" to bend metal, wondering if they managed to psychically produce heat, magnetism, ultraviolet radiation, ionizing radiation, and more. Nothing turned up. But then this seems a bizarrely complex series of causes when he had not eliminated the much simpler one that the children were bending the pieces of metal by hand. It's as if you were trying to work out why the food that you left in your dog's bowl each day disappeared by testing to see if it was struck by lightning and vaporized. No, the dog ate it.

There is a very old principle in science called Occam's razor, named after the thirteenth-century Franciscan friar and philosopher William of Ockham. The original version of this was approximately "Entities must not be multiplied unnecessarily" (it would have been written in Latin), but the gist of the principle is that we should adopt the simplest available theory that explains the results; we should not bring in more complexity than we can help. Occam's razor is clearly not a magic bullet that will solve all problems in science. Sometimes in physics, for instance, the best explanation is by no means the simplest. But the razor is always worth considering. It was not necessary for Taylor to invoke these complex mechanisms involving radiation or electromagnetism until it had been established that there weren't more straightforward reasons for the metal to bend.

Despite, no doubt, being aware of Occam's razor, Taylor continued to postulate that the "Geller effect" was produced by different types of radiation and finally concluded, thanks to some fairly abstruse reasoning, that it was the result of a form of electromagnetism.

When Taylor was carefully watching his child stars and nothing happened, he believed he was observing what he referred to as the "shyness effect." The phenomena he was studying did not take place when the subjects were being observed. Instead, the pieces of metal that were undergoing bending played their part only when the young people involved were out of his sight. The naive interpretation Taylor made was that the ability ceased to function when under direct observation. As Taylor wrote, they "failed to [demonstrate their powers] when prying cameras or scientific measuring instruments proved too much for their self-confidence."

But another group of researchers, working with Taylor's young protégés, managed to get the phenomena to occur while the subjects were being watched, a reality that crushes the possibility of a shyness effect. It seemed rather that Taylor's child stars failed to demonstrate their powers when prying cameras or scientific measuring instruments proved too much for their ability to cheat.

The team from Bath University who took on the young Geller wannabes resorted to a little trickery. They set up the experiment as usual, but in a room where one wall was a one-way mirror. When the observer in the room stopped keeping careful watch, items were indeed bent. But film taken by the researchers behind the mirror made it clear that this happened because as soon as they got the chance, the children were bending pieces of metal with both hands, putting things under the table to bend them, and even putting a rod underneath a shoe to use it to produce a bend.

Just as interesting as the Bath experiment in providing an insight into Professor Taylor's ability to make sensible judgments of how his young protégés were performing their feats is a visit made to him by James Randi in 1975. Randi arrived at Taylor's office at the University of London incognito, in the guise of a journalist. Taylor showed Randi a metal rod with a complex series of bends. When Randi asked how the bending had taken place, Taylor had to admit that he had let the subject, one of his top metal benders, take the metal rod to his (the metal bender's) room. He then returned some while later with the bent rod. This was the closest Taylor came to seeing the bending happen.

What was most revealing was when Taylor showed Randi one of the tamperproof tubes he used to ensure that the children did not manually bend a piece of metal when they took it home or into a different room. It consisted of a transparent plastic tube with a typical chemist's red rubber stopper inserted into each end. To prevent the stopper's being removed a screw had been placed through the plastic tube into the stopper, and then the head of the screw had been sealed with black wax so that it could not be undone without its being noticed.

Inside the tube was one of Taylor's carefully prepared aluminum strips. But this example had been in the hands of one of his metal benders, and the strip had somehow been twisted twice, making a flattened S shape. If this had been done without removing the strip from the tube it was quite remarkable, a feat far beyond a ship in a bottle, requiring the mental influence to cause the metal to practically convert to liquid to allow such dramatic folding in place.

To show just how careful he was with his precautions, Taylor pointed out that the wax had secret markings that meant it could not be replaced without him knowing. This was a case of self-misdirection. Taylor was fooling himself. It turned out that the protection to avoid the wax's being replaced was entirely unnecessary.

When Randi looked closer at the tube he noticed that while one of the stoppers was clearly in good contact with the plastic of the tube, the other seemed not to be well-seated. As he apparently concentrated on the bent strip of metal, Randi used the concealment of his hand to see just how tightly the stopper was held in place—and to his amazement it came right out. There was enough flexibility in the rubber of the stopper to allow it to slip under the screw without the screw moving. Anyone intent on cheating could simply pull the stopper out, remove the strip, bend it, and replace it, then push the stopper back. The tube was totally insecure.

What this emphasizes it not just the poor control conditions applied by Professor Taylor but also his remarkable lack of skepticism. He was totally convinced by his metal benders' ability, even though they performed out of his sight, often taking the tubes home for days at a time, where they encountered such low-grade protection against tampering.

It would be difficult to say just why a scientist would be so lacking in awareness of basic controls. Admittedly, Taylor was a mathematician rather than an experimental scientist, and mathematicians are often considered unworldly individuals. But to understand Taylor's failing, we have to accept that many enthusiasts for a cause (and worst of all, perhaps, converts to a cause) are very good at ignoring evidence that goes counter to their beliefs of what is happening. This shouldn't be the case with scientists, but it often is when the phenomena studied are outside their areas of expertise.

There is a useful parallel in a book I have just received to review. It's a book of UFO photos from across the decades, complete with commentary, compiled by a UFO devotee. I have no ax to grind about the existence of UFOs. I am sure they do exist in the sense of being unidentified flying objects, though because of the impossibly large scale of the universe, I very much doubt that they are extraterrestrial craft. But I am afraid I found many of the comments in the book naive in the extreme. I have no reason to believe that the author is deliberately deceiving the reader—and yet what he has written bears very little resemblance to logical analysis.

To a healthily skeptical eye, many of the pictures look fake. As I mentioned earlier, I went through a phase of mocking up flying saucer pictures in my teens. I didn't do this for any personal gain—I never sent them to newspapers or published them in any way; I simply did it for the fun of it. I used two techniques. Some were plastic models, suspended from near-invisible fishing line and sufficiently out of focus to conceal the fact that they were models. Others were pictures of a metal hubcap, thrown into the air and photographed, spinning, in flight.

One of the problems with the hubcap was that it tended to fly, and so to be photographed, at an unnatural angle. Yet time after time these "unexplained and inexplicable" shots in the book are of fuzzy, out-of-focus hubcaplike objects at the same kind of angle that I found so irritating when I tried to fake my pictures. Two other shots look just like old-fashioned outdoor suspended electric lights with the cable either out of shot or retouched out of the image. Others look like nothing more than a clay pigeon, a bird, or the sun. It's only the way the text describes the shape as having UFO-like characteristics that makes us see a flying saucer.

Worst of all, the book includes the "classic" photo of a collection of UFOs over the Capitol in Washington, D.C. There is nothing in the text to suggest that this picture was debunked long ago. And yet if you take a look at the whole photo, rather than at the cropped image in the book showing just the building's dome and the nearby sky, it is entirely obvious what has happened. The UFO formation is an identical mirroring of a formation of lights running in front of the building. It's nothing more than lens flare, absolutely, definitively. Yet there is not a word about this in the book.

It doesn't take someone setting out to consciously deceive us to present apparently remarkable information that by omission, selection, and blindness to the realities hides the fact that there is a perfectly normal and reasonable explanation.

My intention in this chapter was not to show that all paranormal events are faked or misunderstood. But what it does demonstrate is that the magicians of this world like James Randi who insist on rigid controls before they give any credence to a psi phenomenon are right to do so. Unless we have such careful monitoring and prevention of misdirection, mistake, and fraud, there is no value at all in the information gathered.

So with this timely reminder that some people find it easy to deceive, and others to be deceived, what final conclusions can we draw? Is there anything to the psi phenomena?

11.

IS ANYTHING THERE?

After taking our detailed tour of the history of psi research, it is impossible to say with 100 percent confidence whether or not there are real phenomena being observed, but this is not as much of a problem as it seems. Science is not about proven, definitive facts. Many are shocked by this, but science can never provide us with the absolute truth. Even apparently solid scientific "facts" like the big bang are simply our best-supported interpretation of the current data. It is always possible for new observations and experiments to demonstrate that a cherished aspect of science is wrong. It has happened many times in the past, and it will happen again with some of the theories we now hold.

This is not a bad thing. It is how our scientific understanding evolves. Nor does it suggest, as some might say, that all possible theories should be given equal weight because "science can't prove anything." Some theories have a much higher probability of being correct than others. The widely accepted scientific picture of the universe is based on our current best guesses. Until other data comes along that makes us change our mind, to go with anything other than the best interpretation of the data is simply silly. Why would you choose to go for a less likely option?

It seems probable from all the work that has been done that some aspects of parapsychology don't hold up to scientific testing, although I think others—telepathy particularly—still show distinct possibilities for having a basis in reality. It still might seem odd, despite all that we have experienced in the previous chapters, to dismiss a lot of the evidence for psi phenomena. So many people claim to have experienced strange things happening. You may well have experienced something yourself. Surely there must be something there? Speak to any enthusiast for psychic abilities and he will ask how science can dismiss so many experiences of so many people in pretty well every culture around the globe throughout all of history. After all, he will argue, there's no smoke without fire.

Unfortunately, the real world is very different from the aphorism. (And in reality there is often smoke without fire.) Those who rely on the abundant anecdotal evidence for psi phenomena need to face up to three potential problems: failings of perception, the nature of memory, and the ability of people to cheat. This is why scientists are so wary of anecdotes about psi occurrences—such stories tell us much more about people and their beliefs than they do about actual things that have happened. As Robert L. Park says in his book *Voodoo Science*, "data is not the plural of anecdote."

If we are looking for something that runs counter to the known laws of nature—which is by definition the case with the paranormal—we have to weigh the chance that there is something totally outside of our scientific understanding, which is entirely possible but doesn't happen very frequently, against the chance that someone has made a mistake or has failed to tell the truth—which happens all the time.

It is ironic that when examining extra sensory perception, we are so dependent on our familiar senses, particularly sight, when it comes to making observations. This isn't necessarily true of controlled laboratory tests, but in most anecdotal evidence supporting the existence of psi, we rely on what somebody saw (or, worse, what somebody heard that someone else had seen). Unfortunately, our senses can fool us and do so all the time.

The ancient Greeks were well aware of this. It's why they put such little weight on actually checking things and made decisions based on a comparison of arguments rather than on experiment. In practice this proved a very bad way to do science, leading to such travesties as Aristotle's famous assertion that women had fewer teeth than men, based on his opinions about women rather than on any attempt to count teeth. But we do always need to be aware of the limitations of our senses.

Sight, particularly, is susceptible to being fooled as a matter of course. As we have discussed, (see page 113), we think of sight as working like a video camera, whereas in reality the "image" we view is a construct that the brain assembles from different detection modules. This means that if we see something happening under difficult circumstances—when light is low, for example, or when someone is going out of his way to distract us—it is very easy for our eyes to fool us. We can easily see things that just don't happen.

It's only necessary to take a look at the history of astronomy to realize how easy it is to see what you want or expect to see, rather than what is actually there. Percival Lowell, founder of the Lowell Observatory in Flagstaff, Arizona, was a businessman trained in mathematics with a fascination for astronomy who spent over twenty years full time on astronomical pursuits. Lowell made detailed studies of the surface of Mars and, inspired by the ideas of Italian astronomer Giovanni Schiaparelli, drew maps of the complex system of canals he saw on the surface. This was taken to be clear evidence that there was once an advanced civilization on Mars.

The canals that Lowell so carefully noted and drew do not exist. Although there are a small number of geological structures that could be misinterpreted this way, in most cases there is nothing whatsoever to see. What's more, Lowell's idea that there might be canals was inspired by a mistranslation. Schiaparelli had written about (natural) *channels*, not canals that had been constructed by intelligent beings. Enthusiastic supporters of the existence of psi effects are likely to have a similar inclination to Lowell's to see what they want to see, rather than what their eyes are really telling them.

And then there's memory. Memory, in many ways, is even worse than basic perception as a tool for collecting scientific data. It's strange, when you think about it, that our memories are so poor, because as human beings we are in large part the sum of our memories. Take them away, as happens with some degenerative brain diseases, and what is left no longer seems to be the same person. Memories shape us and our understanding of the world. But memories are at best flexible and are sometimes downright inaccurate.

This is almost inevitable from the way we build memory. It becomes increasingly difficult over time to be sure whether a memory really exists or whether what is recollected is a photograph or a video or an often-told tale. Unlike a carefully noted scientific record, our memories are selective about extremes. The everyday constants of life pass us by without registering, but the extremes get lodged in memory—so we might remember a long hot summer stretching for weeks and weeks in our childhood when really there were only a handful of good days. Conversely, our memories tend to give more weight to very recent experiences, so we might have a string of excellent meals on a vacation, but then think the catering was terrible because the last one was bad.

The other problem with memory is that we don't remember what we saw, but rather what we thought we saw, and that impression gets turned into fact by the memory process. We know it happened—we can remember it. This might be expected to occur with long-term memories. Everyone knows that things become a little fuzzy over the years. But the remarkable thing is that memories are likely to be faulty immediately after an event.

Here's a little memory experiment that you can undertake right now. Put the book down, get hold of a pad and a pen or a pencil, and draw a clock face with Roman numerals on it rather than ordinary Arabic numerals. If there is one in sight, don't look at it. This isn't a trick of any kind—there is no deception—just a chance to test out the nature of memory.

Please don't continue until you have drawn your clock face with Roman numerals. If you are feeling lazy, it's enough to fill in the first six numbers. Should you read any further without doing the drawing you will not be able to complete the experiment.

Look at the number that you have put on the face between 3 (III) and 5 (V). What have you drawn? The vast majority of people will have put IV, which is indeed the usual way to write the number 4 in Roman numerals. However, the convention on clock faces is to use IIII instead of IV. If you have any clocks or watches with Roman numerals, the chances are that this is what you will see in the 4 position. All your life you have been looking at clocks and watches with IIII representing four o'clock. But it has never registered. When this experiment was done with individuals who were asked to look at a clock face for a period of time immediately before the test, they still made the same mistake. Can you really put too much faith in a memory of an apparent psi phenomenon, given our powerful ability to substitute what we expect to see for what's really there?

Here's another example of memories of an event proving faulty immediately afterward, in a case that is considerably more dramatic.

On December 4, 1901, there was a horrendous incident during a seminar on criminology at the University of Berlin. As Professor Franz von Liszt gave his lecture, one of the students interrupted to give an alternative viewpoint to the professor's "from Christian morality." A second student jumped up and disagreed profoundly. He said that he was fed up with these Christian morality arguments. The first student was incensed. He pushed his desk over and strode toward his opponent, pulling a gun from under his coat. There was a fight, the two students wrestling for control of the gun until it went off. The second student fell to the floor, apparently dead.

Not surprisingly, the rest of the class was in shock. Von Liszt picked up the gun and asked for attention. He apologized, telling them that he had staged the event in order to perform an experiment. He now wanted everyone present to write down exactly what they had seen. Still shaken, they all obediently wrote out witness statements. And here's where it gets interesting. The versions that the students gave differed wildly. This was no distant memory and featured no ordinary everyday event. They were giving their recollection of something amazing that had been seared on their memories just minutes before.

When the different reports were compared there were, for example, *eight* different names given for the person who started the fight. Among the observers there were wildly differing accounts of the duration of the event, the order in which things happened, and how the whole scene finished with von Liszt's explanation. Some were convinced that the gunman had run from the lecture room—which he hadn't. He had remained standing over the body.

The point von Liszt hoped to make—and in which he was successful far beyond even his own expectations—was to show just how unreliable witnesses are when giving evidence in court. And it is totally bizarre that we still place so much faith in witness evidence in trials today, given the clear example of this and many other similar psychology experiments since. Witnesses are terrible at getting the facts right. They really aren't good enough to rely on in court. Interestingly, von Liszt found that the inaccuracies were worst when describing the events that were most dramatic—those, for example, involving the gun. It's as if the unexpected nature of the event makes us particularly bad at recalling exactly what happened.

There is an exact parallel with those who remember seeing remarkable events like spoons apparently bending on their own, with no one touching them, under the influence of performers like Uri Geller—again, dramatic events that they believe they have witnessed. There is very good evidence that we can't believe what people "have seen with their own eyes" if the only information we have is their testimony. Without clear video evidence, for example, such recollections are practically worthless.

Just how bad we are at recounting what we have (and haven't) seen is demonstrated wonderfully by research done by Professor Dan Simons at the University of Illinois at Champaign-Urbana. If you haven't seen this, you might like to try out his experiment before reading further. Go to www.universeinsideyou.com and click on "Experiments" and then "Experiment 9—Counting the Passes." Follow the instructions on the screen.

Don't continue reading if you want to try the experiment, or you will reach a spoiler.

The remarkable thing here is that at least half of those taking part in this experiment will fail to accurately recount what they have seen. All they are asked to do is watch a short video, yet they will miss a major part of what has happened. Simply by concentrating on one aspect of the action, it is very easy not to register other, very dramatic things that are taking place. In the video, some students are playing with a basketball and the viewer is asked to count the number of passes of the ball between players wearing a particular color. Afterward, the viewer is asked to report anything strange that was seen during the video.

Around half of those viewing the video will not see that someone dressed in a gorilla suit came on screen, beat his chest in the middle of the picture, and walked off. Even fewer will spot various other incidents that occur during the action. The gorilla is not some minor event in the corner of the screen. It is very obvious and takes center stage. If it is possible to miss something as dramatic as this (and I have seen this happen over and over again when this video is run), it can hardly be surprising that witnesses get it wrong when, with powerful misdirection, they believe they have see something during a psi demonstration that differed from reality.

A particularly interesting study in memory after the event specifically focused on how different people responded to apparent psi phenomena. In this experiment, convincing trickery was used to reproduce popular effects such as bending keys. A group of observers were tested on their recall of what they had seen. Some were skeptical about psi; others were convinced that psychic phenomena were real.

There were significant differences in how the two types of individuals reported what they had witnessed. The skeptics were less likely to rate what they had seen as being paranormal—hardly surprising under the circumstances. But what was particularly interesting was that the members of the believers group were significantly more likely to make errors or omissions in their report of what they had seen. This was particularly the case with information that was crucial to making cheating possible. They would not notice, for example, that a key went briefly out of sight before it was bent. There is no evidence that this was conscious concealment of the facts—instead, their brains edited out what they didn't want to see.

In a second, even more significant experiment, participants witnessed two tests of ESP based on card guessing, in a format similar to the Rhine experiments. In the first test, unknown to the participants, the cards were marked to enable those doing the guessing to cheat. They got a hit rate of three out of five, rather than the expected one out of five that would be the chance result from Zener cards. In the second test there was no cheating and the result was as predicted by chance.

The participants watching the event were again a mix of skeptics and believers. Those who doubted that psi was possible reported what had happened accurately, even though they were uncomfortable with what they had seen in the first run. Many of those who were enthusiasts for the paranormal reported that *both* tests were successful. They remembered what they wanted to remember and wiped out the rest. This kind of recall effect should have no influence on properly recorded scientific tests, but it certainly makes much of the anecdotal evidence open to question. For that matter, there have been many tests over the years, including the early section of Rhine's research, where results weren't recorded until sometime after the test, and such a memory bias could still have had an influence.

The outcome of both the experiments comparing skeptics and believers was that the believers remembered what they had seen in a way that reinforced their beliefs. They saw what they wanted to see. In different circumstances the same effect can result in skeptics missing an actual occurrence because they don't believe it possible, but in a typical psi experiment the biggest danger is that those who want the experiment to be a success (which will usually be the case with the experimenters) will be more likely to miss crucial aspects of what happened.

All this is purely unconscious without any intent to deceive. But there is, of course, also cheating. This is the reason, as we have seen, that so many excellent scientists have been taken in by faked psi phenomena. Of course, not every scientist involved in parapsychology is a good scientist. Some have shown scant regard for scientific protocols, allowing their own beliefs to color their work to the extent that it is worthless. But others, particularly those who have come to parapsychology from fields like physics, have set out with the intent of making genuine tests, only to be easily fooled by fraudsters.

As stage magicians like James Randi and Derren Brown like

to point out, it is much easier to pull the wool over the eyes of a physicist than of a conjurer. A physicist is used to undertaking experiments with materials that have no ulterior motive. If you are studying a photon of light, it may behave in a weird quantum way, but it doesn't attempt to fool you. When it comes to the techniques adopted by the stage experts, it is hardly surprising that it has been relatively easy to fool scientists, even those who think that they are imposing controls.

In looking over the evidence for psi phenomena it might seem surprising how much reference is still made, for example, to Rhine's work in the 1930s or to the likelihood that Uri Geller's work at SRI in the 1970s was genuine. Arguably one of the reasons for this is that most of the scientific work attempting to establish well-controlled evidence for telepathy or remote viewing or telekinesis in recent years has had negative outcomes or has relied on very small statistical effects. We need to look back to earlier work to find consistent sizable positive results, and that itself has to be taken as a clear indicator.

When I am writing about physics, I will certainly frequently refer to the work of the physicists of the 1930s, both to fill in essential parts of the history of science and because there are significant theories from this period that have stood the test of time. But it would be unusual if there had been no good experimental evidence since that had helped clarify any uncertainties that were present at the time—we should expect to see more clear results now, not fewer.

Part of the problem in providing clear proof of psi abilities, as we have seen, is that it is surprisingly difficult to design an experiment that is immune to cheating. Not impossible, but certainly beyond many of those who have undertaken experiments in ESP (Rhine springs to mind). What's more, there is a clear need not only to prevent the person being tested from cheating, but also to design an experiment that minimizes the possibility of the experimenter cheating.

We saw this with the psychic eggs at the Rhine lab, and the need was demonstrated again thanks to some remarkable detective work on the records of Rhine's UK counterpart, Dr. S. G. Soal. For the first few years of his work Soal found very little to report, discovering no one with psi abilities. But in a very long series of tests on Basil Shackleton, a self-proclaimed psychic, Soal was to find what appeared to be conclusive proof of telepathic ability. Soal's experiments were like an advanced version of the Pratt-Woodruff experiment (see page 182), but with the two participants in separate rooms and monitored to avoid cheating. The tests really did seem to be perfectly controlled, and even after years of analysis it is hard to see how Shackleton could have cheated. But remarkably, decades afterward, a mathematician managed to prove that Soal himself had doctored the results.

Rather than rely on the erratic randomization of shuffling a pack of cards, Soal had been using near-random numbers taken from a table of logarithms using a short series of instructions (along the lines of "take the eighth digit of the 100th value," etc.) to generate an acceptable set of unbiased numbers. But statistician Betty Markwick managed to reverse engineer Soal's selection method and re-create the original list of random numbers. What she found was that there were spaces left in the sequence every few guesses. And these spaces aligned with successful guesses far too often to be a coincidence. It seems that Soal was filling in some of the successful guesses after the event.

It might seem terrible that we can't trust the scientist undertaking the experiment. In principle any scientist could doctor his results—and some have been caught doing so. But the temptation seems to be significantly greater than usual with parapsychology, and it is essential that experimental design minimizes the opportunity for the experimenter to cheat, and ensures that any successful experiment is duplicated, with the same results produced by different researchers.

Time and again what we have seen in attempts at controlled psi experiments in universities is that scientists alone, particularly those who weren't psychologists, were highly unsuited to the task because they were not good enough at establishing controls. However many times they were told, they did not seem capable of understanding that their subjects could be cheating.

Perhaps the definitive example of this was the Alpha Experiment. This took place for two years beginning in 1979 at Washington University in St. Louis, Missouri. A laboratory had been set up with a $500,000 grant from James S. McDonnell, then chairman of aerospace giant McDonnell Douglas. The experiments were undertaken by a group led by Peter Philips, a physicist, who like several others before him decided that young people's attempts at spoon bending would be worthy of study.

The reason this experiment is so valuable as a demonstration of how *not* to go about such research is the contribution made by James Randi. The magician and arch skeptic offered his services to spot cheating, and even provided the experimenters with a protocol that should minimize the chances of cheating happening. Both his offer and his protocol were ignored. This was unfortunate for the experimenters, as Randi then arranged for two teenage amateur stage magicians to apply as subjects. They were so far ahead of the other applicants in their apparent abilities that these two plants, Michael Edwards and Steve Shaw, were chosen to be the main subjects of the experiment.

Despite being instructed by Randi to confess if ever directly confronted by the experimenters, the pair undertook trial after trail and amazed Philips and his team. Even though Randi had provided him with advice on appropriate protocols, Philips omitted the most basic controls. So, for example, the pair would be presented with a whole pile of spoons which had been premeasured. Each spoon was carefully labeled, yet after a little manipulation by the two boys, some of the spoons were found to have been bent.

Although some classic spoon bending was performed by hand under the table where possible, the controls were so poor that the boys could appear to produce an effect without ever manipulating a spoon. The labels on the spoons were pieces of paper, attached by a loop of string. The boys would first remove the labels from several spoons, which they said was necessary because the labels got in the way of their ability to focus. They would then make their attempts at bending and replace the labels. All it took was to switch the labels on two similar spoons and hey presto, each had changed shape compared with its original measurements—no two spoons are going to be perfectly identical. This was spoon bending that worked all on its own.

The researchers would later argue, as Rhine did long before them, that there was a distinction between early "exploratory" experiments, where controls were light, used to get a feel for what was possible, and later formal experiments, where controls were tightened. It is certainly true that in the Alpha program later experiments did have stronger controls and produced fewer and fewer results. However, there are two problems with this defense.

First, unless the exploratory experiments are discarded entirely and not reported in any way, the result is to provide a confirmation, particularly for the press, that something real is happening. After all, this was a test by real scientists in a real lab. If it is really necessary to have exploratory experiments, then the results should not be published; the experimenters should mere state that the tests were undertaken. However, there is a huge danger that these early, loosely controlled tests will convince the experimenters themselves of the reality of what they are seeing. Once we establish a strong belief, it is difficult to shake it, and the tendency is to come up with justifications for why later, contradictory evidence occurs, rather than accepting it as making the belief untenable. In the end, scientists are human beings and subject to the usual, irrational aspects of human behavior.

The second problem with having these informal experiments is that data leakage can occur from the exploratory into the formal. If the exploratory results were positive, it is very tempting for researchers to include them with the other, more formal results—and in some experiments it is hard after the fact to establish just where the borderline between exploratory and formal was. What no one has convincingly argued, though, is why there needed to be any exploratory experiments. It's not as if the Alpha researchers weren't aware of previous psi experiments—they knew what to look for. There is no justification for not using full controls and formal techniques from day one.

The Alpha experiments weren't the only ones where rationalization after the fact was employed to try to push negative results out of the way—in fact, this is a common occurrence in psi investigations. I have deliberately avoided until now a number of excuses you will find littering the literature on ESP to provide reasons why a particular experiment "failed" and should be discounted (where failure is confirmation of the "null hypothesis" that a psi ability does not exist). I have ignored these because they seem poor excuses to try to explain away negative results and don't have any place in a proper scientific experiment.

However, I ought to briefly mention these attempts to worm out of the proper conclusions, as these excuses will inevitably come up in reports on psi experiments. The first is the sheep and goats effect. This suggests that psi, being a sensitive cerebral activity, can be influenced by the mental attitude of those present. If onlookers are skeptics (goats), they will tend to suppress the ability of those being tested, while an audience of believers (sheep) will encourage things to happen. This means, of course, that supporters of the sheep and goat effect will make a fervent attempt to exclude from any demonstrations anyone who goes out of his or her way to be vocally skeptical.

This seems the worst sort of rationalization. A much more reasonable conclusion is that psi abilities disappear when the likes of James Randi are around, because a cheating subject knows he or she is likely to be spotted by someone with a magician's expertise. Of course, it is entirely possible that an individual performing a psi test would do worse if she believed herself incapable of doing it—the "Pygmalion effect," where we perform better if we believe we are good at something, is well established in many activities. But to suggest that having nonbelieving observers present can stop experiments from working is simply an excuse. It is interesting that those who try to exclude Randi because their psi abilities will dry up in his presence never seem to have a problem if he turns up in disguise.

A second, more subtle problem that is often identified is the psi experimenter effect. The suggestion here is that where with, say, a basic physics experiment we would expect any competent scientist to get the same result, with a parapsychology experiment, some scientists will be unable to replicate positive results because they "jinx" the experiment. It seems much more likely, though, that an inability to replicate—at least if widespread—reflects problems with the original experiment, rather than with the later experimenters.

In any experiment a replication can fail—and one or two famous theoretical scientists, most notably the physicist Wolfgang Pauli, have had a reputation that they make experiments go wrong if they walk into the room—but this is a matter of scientists' humor rather than a real effect, and with enough experiments, properly replicated, this problem goes away. There seems no reason for considering the psi experimenter effect to have any validity. Taken to the extreme, it totally invalidates the point of experimentation anyway, as positive results are taken to demonstrate psi, while negative results are thought to demonstrate a psi effect of the experimenter suppressing the ability of the subject. There would be no evidence gained from experimenting, because whatever you do establishes the existence of psi. The idea of the psi experimenter effect seems to add an unacceptable layer of complexity to the process.

Psi enthusiasts will tell you that the erratic and unpredictable nature of their experiments and the difficulty in reproducing ESP reflects the very human aspect of what is being tested—but it is more likely simply to reflect that in these particular experiments, nothing at all is being detected. Getting occasional, unpredictable positive results is exactly what we would expect when there are chance coincidences, requiring no psi explanation. I should stress I am not dismissing psi, merely saying that a particular kind of evidence from certain experiments suggests strongly that there is nothing more than random chance at work in these tests.

One other oddity of psi scientific recording is particularly dangerous and has sometimes significantly muddied the waters. Generally speaking, what scientists do in an experiment is test a hypothesis—in the case of ESP tests, they may be trying to establish, for example, if an individual can communicate a piece of information mentally. Because of the statistical methods used, it is often possible with a psi test to turn a bad result into an apparently good one. Imagine, for example, someone is predicting a series of coin tosses, as I did on page 148, but imagine that the sequence is allowed to run much longer, for, say, 1,000 tosses.

On average, with no special abilities, we would expect our subject to guess around 500 values correctly. We would suspect there was something else at play, whether clairvoyance, cheating, or a badly balanced coin, if someone got 600 right answers. But what if he got only 400? This is just as unlikely statistically as a score of 600. In any other kind of experiment, failing in a big way like this would be put down as a straightforward negative outcome, but many psi researchers see this as a good thing, describing it as "psi missing."

The idea of psi missing is that since a score of 400 is just as unlikely as a score of 600, then it is equally likely to have a cause, and perhaps the mental efforts of the subject actually made the clairvoyance (or telepathy or precognition depending on how the trial is run) suggest the *wrong* value rather than the correct one. This is a very tenuous suggestion. After all, it seems bizarre that someone's mental abilities would produce the reverse of what he was trying to do. It seems much more likely that psi missing identifies experiments where there is a poor basis for accepting the statistical analysis used.

A particularly undesirable aspect of bringing psi missing into the equation is that it will double the potential for random fluctuations to be counted as real events. If we are looking for an effect that pushes my coin toss example over 550 heads, say, there will be a certain number of fluke occurrences during a big enough set of runs. But assuming a fair coin, there will also be around the same number of values under 450, which in an ordinary experiment would simply be counted among the failures. If you allow psi missing these become successes too, doubling the number of apparent (but not actual) successes that will appear by chance in a series of runs. Psi missing moves the goalposts. It shouldn't be allowed as a sensible measure of true ESP at work.

In each of these cases, some but not all psi researchers apply this kind of rationalization to attempt to cover up what really ought to be assessed as failure to discover psi—the null hypothesis. It's understandable. In science, the null hypothesis should be just as valuable as finding what you were looking for, but as human beings we are all disappointed by a failure to get where we hoped to be. If, however, we are to be rational and take a truly objective view of what can be counted as successes, we need to gently dislodge the researchers from their hopes and dreams and leave them with the best facts that we can use.

I would say at the conclusion of our voyage of discovery that some aspects of psi are so well discounted by the evidence that there seems no need for any further research, any more than we need to do research on whether or not Roger Bacon's cockatrice might still flourish in some far-flung corner of the globe.

Telekinesis and, especially, spoon bending seem to have few merits. Future gazing disguised as precognition has serious physical issues in terms of causality, and though advanced waves have been put forward as a mechanism, it is difficult to see the need for further research unless something more concrete can be demonstrated. The experimental results produced by Bem and Radin are interesting and need further investigation, but suggest that we have more to learn about randomness and statistics than about telekinesis.

When this kind of statement is made, parapsychologists, who spend their days attempting to assess psi activities, inevitably get rather defensive. Adrian Parker, for instance, a psychology lecturer at the University of Gothenburg in Sweden, comments that "the implication remains that the accused will certainly get a fair trial before being hung or at least left hanging." He says the fact that many psychologists dismiss the possibility of the existence of psi is "dispiriting, but it is hardly unusual."

To an extent Parker has a point. There are plenty of areas of science where one group doubts the value of the work of others. You have only to look at the sometimes acrimonious remarks exchanged by those who have spent their lives working on string theory in physics and those who think it is not even science (or, in the words of one notable physicist, "not even wrong"). But Parker also does not succeed in persuading. He compares the "hardly unusual" dismissal of psi to there being "still little agreement among experts concerning the basic nature of hypnosis and dreams." From this he concludes that we should give parapsychology the benefit of the doubt. Let's unpack that a little.

No one would doubt that dreams exist—the only argument here is over exactly what is going on in the brain during dreaming and why this occurs. Hypnosis is slightly different. Surprisingly, many professionals *do* doubt that hypnosis exists per se, but would certainly not doubt that there are clear examples of people responding to suggestion, without accepting a particular reason for this behavior. But parapsychology has a further hurdle to overcome. Here, many professionals doubt that there is anything happening at all, suggesting that their colleagues have been pursuing chimeras constructed from statistical errors, chance, and cheating. It is a different situation, and however much it irritates parapsychologists, it is important to be brave and come down heavily on those phenomena such as spoon bending and future gazing that seem to have no realistic basis after so many years of testing.

What we are left with are elements of telepathy and possibly aspects of remote viewing that do seem to have the most merit for studying further. Telepathy not only seems the most likely to have a believable physical explanation; it also has not been fully explored in terms of the influence of the relationship between the individuals or the urgency of the need to communicate. This doesn't make it true, but it does suggest that it needs further exploration. Perhaps the biggest concern with remote viewing is that it has hardly ever been observed in UK tests, to the extent that researchers have traditionally joked that "clairvoyance does not work in the UK." Given that there is no difference between people and conditions for such work from country to country, this suggests that better controls might eliminate the possibility of remote viewing.

I wish I could bring this book to a close with a definitive statement on all the topics covered. Indubitably the biggest problem in doing so is the poor quality of scientific investigation into psi. This derives from two major issues, one or both of which have been present in the vast majority of investigations into ESP. The first problem is poor control. From the earliest days of psychic research in the nineteenth century to well into the 1970s poor control was the norm. You have only to look back at the Uri Geller experiments to see that this wasn't just an issue in Victorian times. Much—probably the vast bulk—of the data produced was worthless because of this.

The second issue, which started with Rhine's experiments and has got more and more intense to the present day, is focusing far too much on small variations from the predictions expected by random chance and relying far too heavily on statistical interpretation of the data. At the extreme we see experiments like PEAR's, where all that was being searched for was a small fluctuation in the reading of an electronic device. This trend totally misses the point of psi research because the lab work has become detached from what it is supposed to be studying. It's as if you tried to understand elephants by looking for oddities in the electrical signals passing through a computer model of neurons.

What the researchers seem to have totally forgotten is that they are attempting to verify the validity of hundreds of years of anecdotal evidence. Is there anything behind tales like my own at the start of chapter 3 of telepathy on a remote Scottish island? If we are ever to get anywhere, experimenters should be looking at properly controlled experiments that deal with actual psi phenomena, not statistical nuances. In my experience, the person I appeared to communicate with through telepathy did not develop a vague quantum shift in his thinking pattern; he received the specific words I was silently yelling in my head. Real-world ESP is not about small statistical variations; it is about clear, specific communication.

So if you want to test for telekinesis, don't try to produce a small deviation from expected probability in the output of an electronic device. That's not what telekinesis is about. Set up an experiment where someone moves a physical object using only his mind. This cannot be about influencing the throw of a die. That always was a totally crazy concept, partly because it reintroduces the statistical element we are trying to avoid, and partly because it is pretty well impossible to imagine, even if you could mentally influence the roll of a die, how you could know what to do to get a particular face to end up on top. We just aren't that good at doing physics in our heads.

Instead, a telekinesis experiment should involve an object, carefully isolated from conventional physical forces like air movement and vibration, which the subject has to move with his or her mind. Fifty years ago this would have been very difficult to do, but we now have a huge amount of expertise in isolating test equipment from external forces. Rigs like the LIGO experiment for detecting gravitational waves have to be excruciatingly good at eliminating other inputs. All you would need is a similar setup (though much smaller—the LIGO experiment is several kilometers long) in which the subject had to move an extremely light object, displace an extremely sensitive balance, or put pressure on a movement detector without any means of physical contact. *That* would be a telekinesis experiment—but PEAR wasted its time on electronic fluctuations.

Similarly, for clairvoyance or remote viewing, forget vague descriptions of locations, which have to be scored by judges. (Judges? This isn't *Psychics Got Talent*; it's supposed to be science.) Dispose of guessing sequences of cards and matching success against statistical expectation. Display a randomly selected sentence from a randomly selected page of a randomly selected book on a concealed computer screen. The subject, who has no means to see the screen, writes down what the sentence is. If she gets it right it's a hit. If she doesn't, it's not. Simple binary judgment. She is either right or she isn't.

Of course, there are potential flaws here—some words, for instance, are more common than others in any language. But in practice the sample size to select from is so large (imagine selecting from every sentence of every book on Google Books, for instance) that over a long series of tests it won't be an issue. And anyway, if this really were a concern it would be easy enough to get around it by using books in a language the subject doesn't speak, so she would not have any expectations of specific words. And remember, it's not good enough to pick up that a sentence has "the" or "a" in it. If we scored for partial sentences this would be an issue, but this is a binary yes or no. You either get the whole sentence right or you fail.

A similar approach could be used with telepathy, although you would have to resort to a fair amount of subterfuge to get around Rhine's concern that what we think is telepathy is actually clairvoyance. One way to do this would be to establish the clairvoyance level initially, then look for any extra successes on top of this. More important than making the distinction Rhine was so concerned about would be trying to duplicate the apparent conditions that encourage telepathy: closeness between participants and urgency of communication.

It might seem these elements would be difficult to incorporate, but like Daryl Bem's experiments it would be possible to do this by modifying classic psychology studies. A good example would be the experiments carried out by Stanley Milgram at Yale in the early 1960s. Milgram's subjects were asked to give electric shocks to another person who was behind a glass screen. Supposedly this was to encourage the other person to learn. Under pressure from the experimenter, the subjects applied larger and larger electric shocks until they were reaching lethal levels. In practice there was no electric shock—the person behind the glass screen was part of the experiment, acting out the effects of being electrocuted.

Milgram's intent, in an experiment run at a time when Nazi war trials were still under way, was to see just how far individuals would break accepted moral boundaries under orders. However, it would be easy to envisage an alternative version that tested for telepathy under urgent pressure. In this, the person behind the glass would be required to type a word on a computer with no prompting as to what the word should be. If he got the word right, he would be rewarded. If he got the word wrong, he would receive an increasingly powerful electrical shock. The actual test subject, sitting on the other side of the glass, would know the randomly selected word in advance, and would attempt to provide that word to the victim by telepathy.

As in the Milgram test, no actual shocks would be needed to make the experiment work. But the subject, unaware of this, would be under extreme pressure to get the word across mentally as he or she saw the other individual suffer more and more. Variants could bring in other factors to see if they influenced telepathic ability. Though the distinction could not be proven, this test would be inclined to test telepathy rather than remote viewing, because only the sender is experiencing pressure.

Tests like this are entirely possible, but as yet, as far as I am aware, they have not been carried out in a formal way. It sometimes seems as if researchers are more interested in carrying out work that ensures that they are able to continue with their careers than with producing definitive results. The design of some experiments suggests that the scientists are worried that a clear test could put an end to their work, so instead they would rather make vast numbers of indecisive experiments so that they can keep turning out the papers.

This sounds cynical, but we have to bear in mind that scientists want to keep the money coming in as much as any other human beings, and there is some evidence from history that one of the reasons we tend to get stuck with incorrect theories for longer than is necessary is because researchers avoid undertaking experiments that could render previous decades of their work worthless.

There are potentially valid, if not scientifically detailed, explanations for the mechanisms of some psi phenomena. There is some evidence that has not been proved worthless. So there remains hope for those who *want* there to be something there. For me, coming at this with an open mind while frankly wishing that ESP did exist, I have to conclude that the existing experiments have demonstrated nothing more than coincidence, artifacts of the experimental design, misunderstanding, and fraud.

It's time to switch off the life support for parapsychology in its present form and get the researchers to bite the bullet and go for the real thing.

NOTES

I.

FEEL THE POWER—SUPERHEROES AND PHYSICS

PAGE 4—Stuart Firestein's assertion that real science begins where the facts run out is from Stuart Firestein, *Ignorance* (Oxford: Oxford University Press, 2012), p. 12.

PAGE 5—Fred Hoyle's likening of those insisting that the big bang was the only theory to consider in cosmology to a flock of geese is in Fred Hoyle, Geoffrey Burbidge, and Jayant Narlikar, *A Different Approach to Cosmology* (Cambridge: Cambridge University Press, 2000), p. 78.

PAGE 5—Harry Collins's remarks about scientists relying on prior expectations when they believe the phenomenon does not exist is referenced in Dean Radin, *The Conscious Universe* (New York: HarperCollins, 1997), p. 261.

PAGE 6—John Bell's assertion that physicists need to be

open-minded given his experience with electrostatics in damp Northern Ireland is quoted in David Kaiser, *How the Hippies Saved Physics* (New York: W. W. Norton, 2011), pp. 167–68.

PAGE 9—Roger Bacon's remarks on basilisks, etc., in his letter often known as *De Mirabile* is Roger Bacon, *Letter Concerning the Marvelous Power of Art and of Nature and Concerning the Nullity of Magic* (Whitefish: Kessinger Publishing, 2005), pp. 21–22.

2.
PARAPSYCHOLOGY—SEPARATING SHEEP AND GOATS

PAGE 10—The quote from Yvette Fielding is taken from the London *Daily Mirror*'s website at http://www.mirror.co.uk/3am/celebrity-news/spooky-truth-tvs-most-haunted-563082.

PAGE 10—Video of the tablecloth moving during the event described, suggesting manual assistance, can be seen on YouTube at http://www.youtube.com/watch?v=1sotbwn8AuA.

PAGE 12—Martin Bojowald's quip that string theory is a theory of everything because everything and anything can happen is from Martin Bojowald, *Once Before Time: A Whole Story of the Universe* (New York: Alfred A. Knopf, 2010), p. 83.

PAGE 15—Information on physical mediumship experiments at Scole, England, is from Grant Solomon, *The Scole Experiment* (Waltham Abbey, UK: Campion Books, 2006).

PAGE 19—Douglas Blackburn's admission of attempting to fool the scientists and description of the techniques used is from the London *Daily News*, September 1, 1911.

PAGE 21—Information on the James Randi Educational Foundation and the $1 million prize from http://www.randi.org.

PAGE 24—Information on the regular human senses from Brian Clegg, *The Universe Inside You* (London: Icon Books, 2012), pp. 163–82.

3.
CAN YOU HEAR ME?

PAGE 32—Brian Josephson's Nobel Prize details are taken from the Nobel Prize organization website, http://www.nobelprize.org/nobel_prizes/physics/laureates/1973/josephson.html.

PAGE 33—Brian Josephson's assertion that his "Nobel reflection" in the 2001 stamp pack was both provocative and serious comes from an interview with the author in October 2004.

PAGE 33—The text of the article by Brian Josephson is taken from the Royal Mail Nobel Prize 100th Anniversary Commemorative Pack, issued on October 2, 2001.

PAGE 34—David Deutsch's comment that Josephson's ideas on telepathy were rubbish appeared in the London *Observer*, September 30, 2001.

PAGE 34—The Royal Mail spokesperson's comments on Josephson's article is from "News," *Nature* 413, 339 (September 27, 2001).

PAGE 34—Brian Josephson's defense of his ideas was on the BBC radio current affairs show *Today*, October 2, 2001. There is a transcript at http://www.tcm.phy.cam.ac.uk/~bdj10/stamps/today.html.

PAGE 40—Information on quantum entanglement from Brian Clegg, *The God Effect* (New York: St. Martin's Press, 2006).

PAGE 40—The EPR paper is A. Einstein, B. Podolsky, and N. Rosen, "Can Quantum Mechanical Description of Physical Reality Be Considered Complete," *Physical Review* 47 (May 15, 1935).

PAGE 41—Einstein's remark about God not playing dice is from a letter to Max Born published in Max Born, *The Born-Einstein Letters* (London: Macmillan, 1971), p. 57.

PAGE 44—The details of Nick Herbert's instantaneous quantum entanglement communicator are in David Kaiser, *How the Hippies Saved Physics* (New York: W. W. Norton, 2011), pp. 209–14.

PAGE 47—Elizabeth Rauscher's suggestion that telepathy takes place across imaginary dimensions is in E. A. Rauscher and R. Targ, "The Speed of Thought: Investigation of a Complex Space-time Metric to Describe Psychic Phenomena," *Journal of Scientific Exploration* 15, no. 3: pp. 331–54.

PAGE 49—Brian Josephson's suggestion that we could be able to pick up on patterns in short stretches of random sequences is from B. D. Josephson and F. Pallikari-Viras, "Biological Utilization of Quantum Nonlocality," *Foundations of Physics* 21; (1991): pp. 197–207.

PAGE 50—More details of Kevin Warwick's implant experiments are available from his website, http://www.kevinwarwick.com/.

PAGE 53—Joseph Rhine's suggestion for a nonphysical mechanism for telepathy and remote viewing is described in Joseph Banks Rhine, *Extra-Sensory Perception* (Hong Kong: Forgotten Books, 2008), p. 214.

PAGE 56—Washington Irving Bishop's life and death as "the first mind reader" is described in Richard Wiseman, *Paranormality* (London: Macmillan, 2011), pp. 234–38.

PAGE 58—The attempts to demonstrate telepathy and remote viewing over several hundred miles are described in Joseph Banks Rhine, *Extra-Sensory Perception* (Hong Kong: Forgotten Books, 2008), pp. 65–66.

PAGE 59—Charles Tart's book claiming remarkable telepathy results was Charles T. Tart, *Learning to Use Extrasensory Perception* (Chicago: University of Chicago Press, 1976).

PAGE 60—An analysis of the problems with Charles Tart's ESP training device is found in Martin Gardner, *Science Good, Bad, and Bogus* (New York: Prometheus Books, 1989), pp. 207–14.

PAGE 64—The second Tart telepathy experiment is described in Charles Tart, John Palmer, and Dana Redington, "Effects of Immediate Feedback on ESP Performance: A Second Study," *Journal of the American Society for Psychical Research* 73 (April 1979): pp. 151–65.

PAGE 64—Details of the Maimonides experiments on psi abilities while dreaming are from Simon J. Sherwood and Chris A. Roe, "A Review of Dream ESP Studies Conducted Since the Maimonides Dream ESP Programme," *Psi Wars* (Exeter: Imprint Academic, 2003), pp. 85–106.

PAGE 65—Details of the linkage between the painting *Descent from the Cross* and Winston Churchill are from Dean Radin, *The Conscious Universe* (New York: HarperCollins, 1997), p. 69.

PAGE 66—More details of the random-picture technique for generating ideas can be found in Brian Clegg and Paul Birch, *Instant Creativity* (London: Kogan Page, 2007), p. 74.

PAGE 68—The summary report on ganzfeld experiments is Daryl J. Bem and Charles Honorton, "Does Psi Exist? Replicable Evidence for an Anomalous Process of Information Transfer," *Psychological Bulletin* 115, no. 1 (1994): pp. 4–18.

PAGE 72—The summary of later ganzfeld trials into the 1990s is Julie Milton and Richard Wiseman, "Does Psi Exist? Lack of Replication of an Anomalous Process of Information Transfer," *Psychological Bulletin* 125, no. 4 (1999): pp. 387–91.

PAGE 75—The quantum telepathy experiment undertaken by Abner Shimony and others is detailed in Joseph Hall, Christopher Kim, Brien McElroy, and Abner Shimony, "Wave-Packet Reduction as a Medium of Communication," *Foundations of Physics* 7 (1977): pp. 759–67.

4.

IT MOVES!

PAGE 82—Sir William Crookes's attempt at telekinesis using a chemical balance is described in E. M. Hansel, *ESP: A Scientific Evaluation* (London: MacGibbon & Kee, 1966), p. 154.

PAGE 83—The results of Frick's trial of telekinesis on dice using a control sequence are described in E. M. Hansel, *ESP: A Scientific Evaluation* (London: MacGibbon & Kee, 1966), pp. 156–57.

PAGE 86—A description of James Hydrick's performance and its foiling by the use of Styrofoam chips is found in Richard Wiseman, *Paranormality* (London: Macmillan, 2011), pp. 96–100.

5.

THINGS TO COME

PAGE 91—More information on advanced waves can be found in Brian Clegg, *How to Build a Time Machine* (New York: St Martin's Press, 2011), pp. 152–57.

PAGE 93—Jean Burns's surprise that precognition works as well with quantum randomness as with classical randomness is in Jean Burns, "What Is Beyond the Edge of the Known World," *Psi Wars* (Exeter: Imprint Academic, 2003), p. 10.

PAGE 94—Research by sleep scientists suggesting 80 percent of dreams focus on negative events is mentioned in Richard Wiseman, *Paranormality* (London: Macmillan, 2011), p. 282.

PAGE 94—The tendency to dream about subjects that we are worried about was highlighted in L. Breger, I. Hunter, and R. W. Lane, *The Effect of Stress on Dreams* (New York: International Universities Press, 1971).

PAGE 96—Details of experiments in precognition at the University of Nevada Las Vegas from Dean Radin, *The Conscious Universe* (New York: HarperCollins, 1997), pp. 125–33.

PAGE 96—Selectivity in Radin's reporting is mentioned in Stanley Jetters, "Anomalous Effects Related to Consciousness," *Psi Wars* (Exeter: Imprint Academic, 2003), pp. 142–43.

PAGE 97—Daryl Bem's paper on precognition using variants of standard psychological tests is Daryl J. Bem, "Feeling the Future: Experimental Evidence for the Anomalous Retroactive Influences on Cognition and Affect," *Journal of Personality and Social Psychology* 100 (2011): pp. 400–25.

PAGE 101—The study showing that participants would unconsciously start to match a pattern in a supposedly random sequence is Harold W. Hake and Ray Hyman, "Perception of the Statistical Structure of a Random Series of Binary Symbols," *Journal of Experimental Psychology* 45, no. 1 (January 1953): pp. 64–74.

PAGE 101—Daryl Bem's response to the suggestion of testing the randomness of the input against always guessing the same curtain or alternating curtains is in an e-mail to the author dated July 15, 2012.

PAGE 106—The replication of Bem's research is described in Bob Holmes, "ESP Evidence Airs Science's Dirty Laundry," *New Scientist*, January 18, 2012.

6.

SEEING ELSEWHERE

PAGE 107—Derren Brown's remote-viewing demonstration at Sedona, Arizona, features in the TV show *Derren Brown: Messiah,* Channel 4 TV, first broadcast January 7, 2005.

PAGE 110—Roger Bacon's description of devices for remote viewing is from Roger Bacon, *Opus Majus,* trans. Robert Belle Burke (Philadelphia: University of Philadelphia Press, 1928; repr. Kessinger Publishing), pp. 581–82.

PAGE 113—The basics of the workings of human vision is taken from Brian Clegg, *The Universe Inside You* (London: Icon Books, 2012), pp. 91–96.

PAGE 116—The ESP training machine experiments at SRI are described in Martin Gardner, *Science Good, Bad, and Bogus* (New York: Prometheus Books, 1989), pp. 77–85.

PAGE 119—The experiment on albino rats where experimenters produced different results depending on whether they were told the rats were bright or dull is described in Robert Rosenthal and Kermit L. Fode, "The Effect of Experimenter Bias on the Performance of the Albino Rat," *Behavioral Science* 8 (1963): 183–89.

PAGE 120—Doubts about the reality of Newton's experiments on light and color are described in John Waller, *Leaps in the Dark: The Forging of Scientific Reputations* (Oxford: Oxford University Press, 2004), pp. 91–112.

PAGE 121—The dubious nature of Eddington's Príncipe measurements is from Michael Brooks, *Free Radicals: The Secret Anarchy of Science* (London: Profile Books, 2011), pp. 66–68.

PAGE 123—The Jupiter remote-viewing experiment is described in James Randi, *Flim-Flam! Psychics, ESP, Unicorns, and Other Delusions* (Falls Church, VA: James Randi Foundation, 1982/Kindle edition), location 1221.

PAGE 125—The description of the Semipalatinsk remote-viewing exercise and the positive report of the Swann remote viewing of Jupiter are from H. E. Puthoff, "CIA-Initiated Remote Viewing Program at Stanford Research Institute," *Journal of Scientific Exploration* 10, no. 1 (1996): pp. 63–76.

PAGE 128—The book *Mind-Reach* mentioned in the text is Harold Puthoff and Russell Targ, *Mind-Reach* (London: Jonathan Cape, 1977).

PAGE 128—Martin Gardner's observation about not knowing if the photographs in *Mind-Reach* were taken after the drawings were made is from Martin Gardner, *Science Good, Bad, and Bogus* (New York: Prometheus Books, 1989), p. 316.

PAGE 135—The Italian dowsing trial with proper double-blind controls is described in James Randi, *Flim-Flam! Psychics, ESP, Unicorns, and Other Delusions* (Falls Church, VA: James Randi Foundation, 1982/Kindle edition), location 5383.

PAGE 137—Details of ley lines and Alfred Watkins are from Alfred Watkins, *The Old Straight Track* (London: Sphere Books, 1974).

7.

IN THE RHINE LAB

PAGE 142—Joseph Rhine's inspiration to study ESP as a result of reading about Oliver Lodge's work on telepathy is mentioned by Rhine in Joseph Banks Rhine, *Extra-Sensory Perception* (Hong Kong: Forgotten Books, 2008), p. 12.

PAGE 143—The Rhines' tutoring in "the indicia of deception" is noted by Dr. Walter Franklin Price in his preface to Joseph Banks Rhine, *Extra-Sensory Perception* (Hong Kong: Forgotten Books, 2008), p. 5.

PAGE 144—Joseph Rhine's assertion that it is easy to be confused about which psi phenomenon you are observing is in Joseph Banks Rhine, *Extra-Sensory Perception* (Hong Kong: Forgotten Books, 2008), pp. 31–32.

PAGE 156—Professor Charlie Broad's assertion that a very low probability of results being chance demonstrates the existence of telepathy is in C. D. Broad, "Discussion: the Experimental Establishment of Telepathic Precognition," *Philosophy* 19, no. 74 (1944): p. 261.

PAGE 156—Rhine's assertion that selection of results is permissible if there is a valid reason for supposing individuals differ in ability is in Joseph Banks Rhine, *Extra-Sensory Perception* (Hong Kong: Forgotten Books, 2008), p. 37.

PAGE 158—The report of manipulation of data by psychologist Dirk Smeesters in 2012 is described in Mark Enserink, "Rotterdam Marketing Psychologist Resigns After University Investigates His Data," *AAAS Science Insider* 25 (June 2012), http://news.sciencemag.org/scienceinsider/2012/06/rotterdam-marketing-psychologist.html.

PAGE 159—Joseph Rhine's first two trials, both resulting in negative results, are described in Joseph Banks Rhine, *Extra-Sensory Perception* (Hong Kong: Forgotten Books, 2008), pp. 54–55.

PAGE 161—The description of the flexibility of controls in Rhine's trials is from Joseph Banks Rhine, *Extra-Sensory Perception* (Hong Kong: Forgotten Books, 2008), p. 67.

PAGE 161—The persuasion of Mr. Lecrone by nonperfect conditions is described in Joseph Banks Rhine, *Extra-Sensory Perception* (Hong Kong: Forgotten Books, 2008), p. 79.

PAGE 162—The description of the initial trials on A. J. Linzmayer is from Joseph Banks Rhine, *Extra-Sensory Perception* (Hong Kong: Forgotten Books, 2008), p. 57.

PAGE 162—Joseph Rhine's assessment of A. J. Linzmayer's character is from Joseph Banks Rhine, *Extra-Sensory Perception* (Hong Kong: Forgotten Books, 2008), p. 83.

PAGE 164—Details of the conditions for Linzmayer's impressive 15-card correct run are from Joseph Banks Rhine, *Extra-Sensory Perception* (Hong Kong: Forgotten Books, 2008), p. 86.

PAGE 166—The study generating apparent ESP from one deck of cards predicting the values of another is described in R. R. Willoughby, "Prerequisites for a Clairvoyance Hypothesis," *Journal of Applied Psychology* 19 (1935): pp. 545–50.

PAGE 166—The study showing that hits in guesses of a Zener card sequence were based on the underrepresentation of repeated values in both cards and guesses is discussed in E. Pöppel, "Signifikanz-Artefakte in der experimentellen Parapsychologie," *Zeitschrift für Parapsychologie und Grenzgebiete der Psychologie* 10 (1967): pp. 63–72.

PAGE 167—Hubert Pearce's $2,500-winning perfect 25 cards in a Rhine experiment is described in Martin Gardner, *Science Good, Bad, and Bogus* (New York: Prometheus Books, 1989), p. 218.

PAGE 168—Joseph Rhine's assessment of Hubert Pearce's personality and honesty is from Joseph Banks Rhine, *Extra-Sensory Perception* (Hong Kong: Forgotten Books, 2008), p. 101.

PAGE 169—The statistics for Hubert Pearce's many series of tests and the description of the working conditions for such tests are from Joseph Banks Rhine, *Extra-Sensory Perception* (Hong Kong: Forgotten Books, 2008), p. 102.

PAGE 174—Hansel's description of his discoveries about the possibility of Pearce cheating on visiting the Duke site is in C. E. M. Hansel, *ESP: A Scientific Evaluation* (London: MacGibbon & Kee, 1966), pp. 74–85.

PAGE 176—The assorted variants of the data given by Rhine and Pratt from the Pearce-Pratt experiment are described in C. E. M. Hansel, *ESP: A Scientific Evaluation* (London: MacGibbon & Kee, 1966), pp. 83–84.

PAGE 176—Rhine's description of remote tests and the use of a telegraph key is in Joseph Banks Rhine, *Extra-Sensory Perception* (Hong Kong: Forgotten Books, 2008), p. 65.

PAGE 180—The experiment involving subjecting an elephant to LSD is described in Alex Boese, *Elephants on Acid and Other Bizarre Experiments* (London: Pan Books, 2009), pp. 111–14.

PAGE 181—Rhine's description of work with George Zirkle is described in Joseph Banks Rhine, *Extra-Sensory Perception* (Hong Kong: Forgotten Books, 2008), p. 65.

PAGE 182—The description of the Pratt-Woodruff experiment and the concerns about the procedure and results are taken from C. E. M. Hansel, *ESP: A Scientific Evaluation* (London: MacGibbon & Kee, 1966), pp. 86–103.

PAGE 184—Details of Levy's misrepresentation of data at the Rhine lab from Thomas Sebeok, "Psychokinetic Fraud," *Scientific American*, September 1974.

PAGE 186—Rhine's assertion that an academic was unlikely to commit long-term fraud is from Joseph Banks Rhine, *Extra-Sensory Perception* (Hong Kong: Forgotten Books, 2008), p. 150.

PAGE 187—The numbers of trials at other universities attempting to reproduce Rhine's effects are from C. E. M. Hansel, *ESP: A Scientific Evaluation* (London: MacGibbon & Kee, 1966), p. 60.

8.

ENTER THE MILITARY

PAGE 189—The account of Major General Stubblebine's attempt to walk through a wall is from Jon Ronson, *The Men Who Stare at Goats* (London: Macmillan, 2005), p. 2.

PAGE 190—The DIA report on Soviet psi activity, prepared by the U. S. Army Office of the Surgeon-General Medical Intelligence Office, John D. LaMothe, *Controlled Offensive Behavior—USSR* is available to download from http://www.dia.mil/public-affairs/foia/pdf/cont_ussr.pdf.

PAGE 191—The quote about Soviet research into parapsychology is from John D. LaMothe, *Controlled Offensive Behavior—USSR,* http://www.dia.mil/public-affairs/foia/pdf/cont_ussr.pdf, p. xi.

PAGE 192—The statement that the Soviet Union was well aware of the benefits and applications of parapsychology research is from John D. LaMothe, *Controlled Offensive Behavior—USSR,* http://www.dia.mil/public-affairs/foia/pdf/cont_ussr.pdf, p. 24.

PAGE 193—Captain LaMothe's proviso on the lack of endorsement for apports and astral projection is from John D. LaMothe, *Controlled Offensive Behavior—USSR,* http://www.dia.mil/public-affairs/foia/pdf/cont_ussr.pdf, p. 27.

PAGE 194—The description of Nina Kulagina's psychokinetic abilities is from John D. LaMothe, *Controlled Offensive*

Behavior—USSR, http://www.dia.mil/public-affairs/foia/pdf/cont_ussr.pdf, pp. 35–36.

PAGE 194—Doubts about the experimental procedure and sizes of data sets in Soviet experiments are expressed in Authors redacted, *Paraphysics R&D—Warsaw Pact,* http://www.dia.mil/public-affairs/foia/pdf/pa_warsaw.pdf, p. 3.

PAGE 195—The visit of a CIA operative to the SRI and the provision of funding is described in David Kaiser, *How the Hippies Saved Physics* (New York: W. W. Norton, 2011), pp. 90–91.

PAGE 196—Guy Savelli's story of killing a goat mentally is in Jon Ronson, *The Men Who Stare at Goats* (London: Macmillan, 2005), pp. 66–73.

PAGE 197—Details of military remote-viewing exercises are from Jon Ronson, *The Men Who Stare at Goats* (London: Macmillan, 2005), pp. 75–125.

9.

PICKING THE PEAR

PAGE 199—The narration of the random mechanical cascade experiment is derived from a photograph on the PEAR website located at http://www.princeton.edu/~pear/images/hm-rmc.jpg.

PAGE 200—The background and history of the PEAR project are from the PEAR website, http://www.princeton.edu/~pear/.

PAGE 201—Analysis and context of PEAR results are from Robert G. Jahn and Brenda J. Dunne, "The PEAR Proposition," *Journal of Scientific Exploration* 19, no. 2 (2005): pp. 195–245.

PAGE 202—Alan Sokal's parody paper taken seriously by the social sciences intelligentsia is Alan Sokal, "Transgressing the Boundaries: Towards a Transformative Hermeneutics of Quantum Gravity," *Social Text* 46/47 (1996): pp. 217–52.

PAGE 203—Alan Sokal's point that he was not just putting down ivory tower academics, but defending science under attack is in Alan Sokal, *Beyond the Hoax: Philosophy, Culture, and Science* (Oxford: Oxford University Press, 2010), p. 329.

PAGE 204—Richard Dawkins's dismissal of the significance of evidence is described in Rupert Sheldrake, *The Science Delusion* (London: Coronet, 2012), pp. 256–57.

PAGE 204—The description of the PEAR team's colleagues' disparagement is from Robert G. Jahn and Brenda J. Dunne, "The PEAR Proposition," *Journal of Scientific Exploration* 19, no. 2 (2005): p. 204.

PAGE 205—Stanley Jeffers's concerns about the analysis of PEAR data are expressed in Stanley Jeffers, "Anomalous Effects Related to Consciousness," *Psi Wars* (Exeter: Imprint Academic, 2003), pp. 148–50.

10.

BENDING SPOONS

PAGE 209—The episode of *The Dimbleby Talk-in* first broadcast on the BBC on the evening of November 23, 1973, is available in a poor recording on YouTube, http://www.youtube.com/watch?v=jTtnQW8f-Oo.

PAGE 211—The assertion that several hundred people telephoned and wrote to the BBC after Geller's broadcast is from John Taylor, *Superminds* (London: Pan Books, 1976), p. 9.

PAGE 212—Basic biographical details of Uri Geller are from his website, site.uri-geller.com.

PAGE 212—The assertion that Uri Geller became aware of his abilities at the age of three is from John Taylor, *Superminds* (London: Pan Books, 1976), p. 14.

PAGE 212—The assertion in *Nature* that extra sensory perception exists is the abstract of Russell Targ and Harold Puthoff, "Information Transmission Under Conditions of Sensory Shielding," *Nature* 251 (October 18, 1974): pp. 602–7.

PAGE 213—John Taylor's account of Geller's TV performance is from John Taylor, *Superminds* (London: Pan Books, 1976), p. 10.

PAGE 214—The example of Uri Geller describing an image inside a sealed envelope with and without the conditions that allowed for cheating is from James Randi, *The Magic of Uri Geller* (New York: Ballantine Books, 1975), pp. 26–31.

PAGE 217—The video evidence of Uri Geller peeking on the *Noel's House Party* TV show during a telepathy exercise is at http://www.youtube.com/watch?v=uRKLvscWeo4.

PAGE 217—The example of James Randi reproducing Uri Geller's telepathy act despite being frustrated in the most obvious ways of cheating can be seen at http://www.youtube.com/watch?v=JPt-7j3ahPo.

PAGE 219—Sherlock Holmes's remark about eliminating the impossible is from Sir Arthur Conan Doyle, *The Sign of Four, in The Penguin Complete Sherlock Holmes* (London: Penguin, 1981), p. 111.

PAGE 222—The revelation that the producer and director of the Dimbleby show later admitted that Geller had been left alone with the cutlery before the show is from James Randi, *The Magic of Uri Geller* (New York: Ballantine Books, 1975), p. 158.

PAGE 224—The strange occurrences during the radio appearance by James Pyczynski are described in James Randi, *The Magic of Uri Geller* (New York: Ballantine Books, 1975), p. 192.

PAGE 226—Donald Singleton's observation of Uri Geller cheating is from Donald Singleton, "How Does Uri Do It," *New York Daily News,* November 5 and 8, 1973.

PAGE 226—The video evidence of Uri Geller bending a key on RAI can be seen at http://www.youtube.com/watch?v=NnDHP OWXFVI.

PAGE 226—Uri Geller's impressive bending of a mystery ladle can be watched at http://www.youtube.com/watch?v=ophFXB nsexw.

PAGE 227—The science claims on Uri Geller's website are in the science section of site.uri-geller.com/en/uri_geller_s_full_biography.

PAGE 228—The editorial expressing reservations about the Targ and Puthoff paper is "Investigating the Paranormal" *Nature* 251 (October 18, 1974): pp. 559–60.

PAGE 229—Information on the SRI experiments and flaws in their controls is from James Randi, *The Magic of Uri Geller* (New York: Ballantine Books, 1975), pp. 45–60.

PAGE 234—The quote from psychologist Charles Rebert stating that Geller cheated is from Joseph Hanlon, "Uri Geller and Science," *New Scientist*, October 17, 1974.

PAGE 234—The quote from astronaut Edgar Mitchell is in James Randi, *The Magic of Uri Geller* (New York: Ballantine Books, 1975), p. 55.

PAGE 234—James Randi's description of Geller's 1973 *Time* magazine demonstration is from James Randi, *The Magic of Uri Geller* (New York: Ballantine Books, 1975), pp. 117–23.

PAGE 235—Andrija Puharich's description of Geller's 1973 *Time* magazine demonstration is quoted by Randi from Andrija Puharich, *Uri: The Journal of the Mystery of Uri Geller* (New York: Anchor Books, 1974), pp. unspecified.

PAGE 236—The description of Uri Geller's encounter with a UFO and the subsequent dematerialization of the film cartridge is from John Taylor, *Superminds* (London: Pan Books, 1976), p. 24.

PAGE 236—Randi's spoon bending for Dr. John Halsted of London University is described in James Randi, *The Magic of Uri Geller* (New York: Ballantine Books, 1975), pp. 169–71.

PAGE 238—Details of the methods used by Geller described by Hannah Shtrang and others were in an article in *Haolam Hazeh*, February 20, 1974, and reproduced in an English translation in James Randi, *The Magic of Uri Geller* (New York: Ballantine Books, 1975), pp. 205–22.

PAGE 241—The suggestion that fraud in Geller's SRI experiments would require "gross collusion" is in John Taylor, *Superminds* (London: Pan Books, 1976), p. 54.

PAGE 242—John Taylor's description of a "rash of spoon-bending all over England" is from John Taylor, *Superminds* (London: Pan Books, 1976), p. 64.

PAGE 243—The description of John Taylor's standardized metal-bending tests is from John Taylor, *Superminds* (London: Pan Books, 1976), p. 78.

PAGE 244—The description of a metal bender breaking a strip of copper when left with it is from John Taylor, *Superminds* (London: Pan Books, 1976), p. 79.

PAGE 246—The Bath University exposure of cheating by children taking part in metal bending is described in Brian Pamplin and H. M. Collins, "Spoon Bending: An Experimental Approach," *Nature* 257 (September 4, 1975): p. 8.

PAGE 246—James Randi's 1975 visit to test John Taylor's abilities to supervise cheating is described in James Randi, *The Magic of Uri Geller* (New York: Ballantine Books, 1975), pp. 159–64.

PAGE 248—The UFO photographs book with the credulous commentary is B. J. Booth, *UFOs Caught on Film* (Newton Abbot: David & Charles, 2012).

11.
IS ANYTHING THERE?

PAGE 252—The quote "data is not the plural of anecdote" is from Robert L. Park, *Voodoo Science* (Oxford: Oxford University Press, 2001), p. 23.

PAGE 255—The experiment recalling a clock face with Roman numerals was used in C. C. French and A. Richards, "Clock This! An Everyday Example of a Schema-Driven Error in Memory," *British Journal of Psychology* 84 (1993): pp. 249–53.

PAGE 256—The inconsistent reporting of the events during the criminology lecture in Berlin are described in Alex Boese, *Electrified Sheep* (London: Pan Books, 2012), pp. 145–48.

PAGE 258—Details of the counting-the-passes/gorilla experiment can be found in Christopher Chabris and Dan Simons, *The Invisible Gorilla* (New York: Crown, 2010).

PAGE 259—The experiment in which believers in psi failed to remember events where cheating occurred is described in R. Wiseman and R. L. Morris, "Recalling Pseudo-Psychic Demonstrations," *British Journal of Psychology* 86 (1995): pp. 113–25.

PAGE 260—The experiment in which believers in psi describe a "failed" attempt as a success is in W. H. Jones and D. Russell, "The Selective Processing of Belief Disconfirming Information," *European Journal of Social Psychology* 10 (1980): pp. 309–12.

PAGE 262—The statistical analysis showing that Soal's results had been doctored by leaving gaps in the random numbers to be filled in is from James Randi, *Flim-Flam! Psychics, ESP, Unicorns, and Other Delusions* (Falls Church: James Randi Foundation 1982/Kindle edition), locations 4059–93.

PAGE 263—The Alpha Experiment and its flaws are described in

James Randi, "The Project Alpha Experiment: Part 1. The First Two Years," *Skeptical Inquirer* 7, no. 4 (1983), available online at http://www.banachek.org/nonflash/project_alpha.htm.

PAGE 270—The suggestion that some scientists are too quick to dismiss parapsychology is from Adrian Parker, "We Ask Does Psi Exist—But Is This the Right Question and Do We Really Want an Answer Anyway," *Psi Wars* (Exeter: Imprint Academic, 2003), pp. 112–13.

INDEX